# SpringerBriefs in Applied Sciences and Technology

SpringerBriefs present concise summaries of cutting-edge research and practical applications across a wide spectrum of fields. Featuring compact volumes of 50 to 125 pages, the series covers a range of content from professional to academic.

Typical publications can be:

- A timely report of state-of-the art methods
- An introduction to or a manual for the application of mathematical or computer techniques
- A bridge between new research results, as published in journal articles
- A snapshot of a hot or emerging topic
- An in-depth case study
- A presentation of core concepts that students must understand in order to make independent contributions

SpringerBriefs are characterized by fast, global electronic dissemination, standard publishing contracts, standardized manuscript preparation and formatting guidelines, and expedited production schedules.

On the one hand, **SpringerBriefs in Applied Sciences and Technology** are devoted to the publication of fundamentals and applications within the different classical engineering disciplines as well as in interdisciplinary fields that recently emerged between these areas. On the other hand, as the boundary separating fundamental research and applied technology is more and more dissolving, this series is particularly open to trans-disciplinary topics between fundamental science and engineering.

Indexed by EI-Compendex, SCOPUS and Springerlink.

More information about this series at http://www.springer.com/series/8884

Jingzheng Ren

# Advanced Operations Management for Complex Systems Analysis

Springer

Jingzheng Ren
Department of Industrial
and Systems Engineering
Hong Kong Polytechnic University
Hong Kong SAR, China

ISSN 2191-530X ISSN 2191-5318 (electronic)
SpringerBriefs in Applied Sciences and Technology
ISBN 978-3-030-45417-3 ISBN 978-3-030-45418-0 (eBook)
https://doi.org/10.1007/978-3-030-45418-0



This Springer imprint is published by the registered company Springer Nature Switzerland AG
The registered company address is: Gewerbestrasse 11, 6330 Cham, Switzerland

# Contents

# Chapter 1
# Advanced Operations Management for Complex Systems Analysis: Introduction

**Abstract** This chapter has had a brief overview of multi-criteria decision making methods for operations management, especially for complex decision-making and complex system analysis, and the main content of each chapter has been introduced: a two-stage interval best-worst method based on the multiplicative constraint was developed in Chap. 2, 2-tupe DEMATEL (decision making trial and evaluation laboratory) was introduced in Chap. 3, fuzzy best-worst network method combined with ISM was proposed for analyzing the complex systems was illustrated in Chap. 4, and a multi-stakeholder intuitionistic fuzzy multi-criteria decision making method was presented in Chap. 5. Typical problems about complex decision-making and complex system analysis were investigated to show the applicability of these advanced operations management methods.

Operations Management represents the countermeasures, strategies and actions used for business practice to achieve the highest efficiency and productivity for an organization, and it aims to transform materials and labor into products and services with the highest efficiency and productivity. The concept of operations management has been popularized and widely used in various fields, i.e. healthcare, construction industry, tourism, aviation engineering and waste management, etc. Operations management usually involves the investigation and analysis of complex decisions or complex systems, because the operations and management of healthcare, construction industry, tourism, aviation engineering and waste management involved various factors and multiple conflicting objectives, and these factors are always independent and interacted. There are various techniques based on mathematical theories and psychology for organizing and analyzing the complex decisions and complex systems. As for analyzing the complex decisions, multi-criteria decision analysis (MCDA) or multi-criteria decision making (MCDM) as a sub-discipline of operation research was widely used for evaluating the alternatives with the considerations of multiple conflicting criteria (Mendoza and Martins 2006; Linkov et al. 2006).

There are various MCDA or MCDM methods, and the most widely used in the past decades is Analytic Hierarchy Process (AHP) developed by Saaty (2008). The most significant advantage of AHP is that it can successfully decompose the complex problem into different hierarchies, thus, it has been used in various complex



J. Ren, *Advanced Operations Management for Complex Systems Analysis*, SpringerBriefs in Applied Sciences and Technology,
https://doi.org/10.1007/978-3-030-45418-0_1

problems including choice, ranking, prioritization, resource allocation, benchmarking and quality management, etc. (Forman and Gass 2001). In order to overcome the problem of vagueness and ambiguity existing in human's judgments, various extended AHP method has been developed by combined with fuzzy set theory, grey theory, interval theory and hesitant fuzzy set theory, and these extend AHP method include fuzzy AHP (Chang 1996), grey AHP (Sahoo et al. 2016), interval AHP (Entani and Sugihara 2012) and hesitant fuzzy AHP (Öztaysi et al. 2015). And it is also difficult for the users to guarantee the consistency among these comparisons. There also some other MCDM or MCDA method such as TOPSIS (Technique for Order Preference by Similarity to an Ideal Solution) (Tzeng and Huang 2011), GRA (grey relational analysis) (Kuo et al. 2008), ELECTRE family methods (ELimination Et Choix Traduisant la REalité, ELimination and Choice Expressing REality) (Figueira et al. 2013), PROMETHEE (Preference Ranking Organization METHod for Enrichment of Evaluations) (Brans and De Smet 2016) and CODAS (Combinative Distance-based Assessment Method) (Keshavarz Ghorabaee et al. 2016), etc. And almost all these MCDM or MCDA methods replies on using AHP or extend AHP methods to determine the weights of the criteria used in multi-criteria decision problems. However, AHP employs the numbers from 1 to 9 and their reciprocals to represents the relative importance/priority of each pair of alternatives/factors. If there are N objects, the users must carry out N(N − 1)/2 times of comparisons when using any of these methods belonging to the AHP family. Moreover, it is difficult for the users to guarantee the consistency among these comparisons when using AHP or extended AHP methods. Accordingly, Rezaei (2015, 2016) innovatively developed a great method, so-called "best-worst method (BWM)", to determine the relative importance/priorities of multiple factors/alternatives. Different from AHP or extended AHP methods, the users can firstly determine the best (i.e., the most important and most preferable) criterion and the worst (i.e., the least important and least preferable) criterion, then compare the best criterion with other criteria and other criteria with the least criterion to establish the best-to-others (BO) and other-to-worst (OW) vectors for determining the weight of each criteria.

BWM has two significant advantages: one is less comparisons with high accuracy, and another convenience for high consistency (Ren et al. 2018). This method has been widely used in many fields for its advantages, i.e., technology selection for urban sewage sludge treatment (Ren et al. 2017), supplier evaluation (Rezaei et al. 2015, 2016), service quality evaluation of airline industry (Gupta 2018), selection of biomass thermochemical conversion technology (van de Kaa et al. 2017), and medical tourism development (Abadi et al. 2018), etc. Based on the work of Rezaei (2015, 2016), BMW method has been improved for more wide application, including interval BMW (Ren et al. 2018), fuzzy BWM (Guo and Zhao 2017), intuitionistic fuzzy multiplicative best-worst method (Mou et al. 2016) and Z-number extended BMW (Aboutorab et al. 2018), etc. Similarly, a two-stage interval best-worst method based on the multiplicative constraint was developed in Chap. 2 of this book to address the vagueness, ambiguity and subjectivity existing in human judgments. In order to illustrate the developed two-stage interval best-worst method, the factors influencing airport competitiveness were prioritized by the developed two-stage interval best-worst method, and implications were also provided based on the obtained results for operations management.

As for analyzing the complex systems with multiple interacted influential factors, decision making trial and evaluation laboratory (DEMATEL), fuzzy cognitive map (FCM) and Interpretative Structural Modeling (ISM) are widely used to investigate the interacted and interdependent factors in the complex systems. DEMATEL technique was developed by the Geneva Research Centre of the Battelle Memorial Institute in 1970s to investigate the complicated causal relationships among multiple factors through matrixes or digraphs (Gabus and Fontela 1972, 1973). The integer numbers including 0, 1, 2, 3 and 4, representing "No influence (0)", "Low influence (1)", "Medium influence (2)", "High influence (3)", and "Very high influence (4)", respectively, were usually used to describe the influence of one factor to another (Ren et al. 2013). It allows multiple stakeholders/decision-makers to express their preferences/opinions and participate in the process of investigating the causal relationships among these factors. The five-scale approach for describing the influence between each pair of factors cannot accurately describe the influence level because of the vagueness and ambiguity existing in human's judgments. Accordingly, many improved DEMATEL methods were developed to fill in this gap by combining with fuzzy set theory, interval theory and interval type-2 fuzzy set theory, etc. Accordingly, Fuzzy DEMATEL (Wu and Lee 2007), grey DEMATEL (Xia et al. 2015), interval type-2 fuzzy DEMATEL (Abdullah and Zulkifli 2015) and some other improved DEMATEL methods were developed. 2-tupe DEMATEL by combining 2-tupe linguistic variables and the traditional DEMATEL method was developed in Chap. 3 of this book, and the cause-effect relationships among the factors influencing sustainable tourism industry.

Besides these, the interpretive structural modelling as a computer-based technique for small groups to develop graphical representations of complex systems was also widely used for investigating the complex systems (Attri et al. 2013). ISM has been widely used in vendor selection (Mandal and Deshmukh 1994), competitiveness improvement of small and medium enterprise (Singh et al. 2007), green supply chain management (Agi and Nishant 2017), lean remanufacturing (Vasanthakumar et al. 2016), six sigma implementation (Alidrisi 2014) and medical tourism development (Debata 2013), etc. ISM can effectively divide the influential factors with complex interactions and interdependences into different levels and identify the cause-effect relationship. For instance, ISM can, prioritize the strategies, identify the most fundamental strategy and divide them into different levels if there are multiple strategies for operations management. However, the prioritization of the factors by ISM was determined based on the driving power and dependence power diagram and ISM based hierarchical model rather than quantitative calculation. Therefore, ISM was usually combined with some other operations management methods. For instance, Thakkar et al. (2005) combined ISM and analytic network process (ANP) for the selection of thirty-party logistics. Similarly, fuzzy best-worst network method combined with ISM was proposed for analyzing the complex systems in Chap. 4, and the combined method was illustrated to investigate the enablers of aviation maintenance safety.

Operations management usually involves multiple stakeholders/decision-makers, and the decisions in the complex systems usually involves the preferences/opinions of multiple stakeholders/decision-makers, and different groups of multiple stakeholders/decision-makers have different preferences/opinions. Accordingly, the so-called "group decision-making" or "multi-actor multi-criteria decision making" methods were developed for complex decision-making with the considerations of the preferences/opinions of multiple stakeholders/decision-makers. Shih et al. (2007) developed an extend TOPSIS method for group decision-making. Sanayei et al. (2010) developed a fuzzy VIKOR (Vlse Kriterijumska Optimizacija Kompromisno Resenje which means multicriteria optimization and compromise solution, in Serbian) method for group decision-making. Ren and Liang (2017) developed a fuzzy multi-criteria decision making method for selecting the most sustainable marine fuel by combining fuzzy logarithmic least squares method and fuzzy TOPSIS. Wang et al. (2018) developed an interval-valued fuzzy multi-criteria decision making method by combining interval-valued fuzzy DEMATEL and interval-valued fuzzy GRA. In Chap. 5, multi-stakeholder intuitionistic fuzzy analytic hierarchy process was employed to determine the weights of the criteria for the evaluation of the alternatives, and multi-stakeholder intuitionistic fuzzy grey relational analysis was used to rank the alternatives. An typical problem about technology selection for waste management which involves the preferences and opinions of multiple groups of stakeholders/decision-makers was studied by the developed interval-valued fuzzy multi-criteria decision making method.

## References

H. Aboutorab, M. Saberi, M.R. Asadabadi, O. Hussain, E. Chang, ZBWM: the Z-number extension of best worst method and its application for supplier development. Expert Syst. Appl. **107**, 115–125 (2018)

L. Abdullah, N. Zulkifli, Integration of fuzzy AHP and interval type-2 fuzzy DEMATEL: an application to human resource management. Expert Syst. Appl. **42**(9), 4397–4409 (2015)

F. Abadi, I. Sahebi, A. Arab, A. Alavi, H. Karachi, Application of best-worst method in evaluation of medical tourism development strategy. Decis. Sci. Lett. **7**(1), 77–86 (2018)

M.A. Agi, R. Nishant, Understanding influential factors on implementing green supply chain management practices: an interpretive structural modelling analysis. J. Environ. Manage. **188**, 351–363 (2017)

H. Alidrisi, Prioritizing critical success factors for six sigma implementation using interpretive structural modeling. Am. J. Ind. Bus. Manage. **4**(12), 697 (2014)

R. Attri, N. Dev, V. Sharma, Interpretive structural modelling (ISM) approach: an overview. Res. J. Manage. Sci. **2**(2), 3–8 (2013)

J.P. Brans, Y. De Smet, PROMETHEE methods, in *Multiple Criteria Decision Analysis* (Springer, New York, 2016), pp. 187–219

D.Y. Chang, Applications of the extent analysis method on fuzzy AHP. Eur. J. Oper. Res. **95**(3), 649–655 (1996)

B.R. Debata, K. Sree, B. Patnaik, S.S. Mahapatra, Evaluating medical tourism enablers with interpretive structural modeling. Benchmarking: Int. J. **20**(6), 716–743 (2013)

T. Entani, K. Sugihara, Uncertainty index based interval assignment by interval AHP. Eur. J. Oper. Res. **219**(2), 379–385 (2012)

J.R. Figueira, S. Greco, B. Roy, R. Słowiński, An overview of ELECTRE methods and their recent extensions. J. Multi-Criteria Decis. Anal. **20**(1–2), 61–85 (2013)

E.H. Forman, S.I. Gass, The analytic hierarchy process—an exposition. Oper. Res. **49**(4), 469–486 (2001)

A. Gabus, E. Fontela, World problems, an invitation to further thought within the framework of DEMATEL (1972)

A. Gabus, E. Fontela, Perceptions of the world problematique: communication procedure, communicating with those bearing collective responsibility (1973)

S. Guo, H. Zhao, Fuzzy best-worst multi-criteria decision-making method and its applications. Knowl.-Based Syst. **121**, 23–31 (2017)

H. Gupta, Evaluating service quality of airline industry using hybrid best worst method and VIKOR. J. Air Transp. Manage. **68**, 35–47 (2018)

M. Keshavarz Ghorabaee, E.K. Zavadskas, Z. Turskis, J. Antucheviciene, A new combinative distance-based assessment (CODAS) method for multi-criteria decision-making. Econ. Comput. Econ. Cybern. Stud. Res. **50**(3) (2016)

Y. Kuo, T. Yang, G.W. Huang, The use of grey relational analysis in solving multiple attribute decision-making problems. Comput. Ind. Eng. **55**(1), 80–93 (2008)

I. Linkov, F.K. Satterstrom, G. Kiker, C. Batchelor, T. Bridges, E. Ferguson, From comparative risk assessment to multi-criteria decision analysis and adaptive management: recent developments and applications. Environ. Int. **32**(8), 1072–1093 (2006)

A. Mandal, S.G. Deshmukh, Vendor selection using interpretive structural modelling (ISM). Int. J. Oper. Prod. Manag. (1994)

G.A. Mendoza, H. Martins, Multi-criteria decision analysis in natural resource management: a critical review of methods and new modelling paradigms. For. Ecol. Manage. **230**(1–3), 1–22 (2006)

Q. Mou, Z. Xu, H. Liao, An intuitionistic fuzzy multiplicative best-worst method for multi-criteria group decision making. Inf. Sci. **374**, 224–239 (2016)

B. Öztaysi, S.Ç. Onar, E. Boltürk, C. Kahraman, Hesitant fuzzy analytic hierarchy process, in *2015 IEEE International Conference on Fuzzy Systems (FUZZ-IEEE)* (IEEE, 2015), pp. 1–7

J. Ren, A. Manzardo, S. Toniolo, A. Scipioni, Sustainability of hydrogen supply chain. Part I: Identification of critical criteria and cause–effect analysis for enhancing the sustainability using DEMATEL. Int. J. Hydrogen Energy **38**(33), 14159–14171 (2013)

J. Ren, H. Liang, Measuring the sustainability of marine fuels: a fuzzy group multi-criteria decision making approach. Transp. Res. Part D: Transp. Environ. **54**, 12–29 (2017)

J. Ren, H. Liang, F.T. Chan, Urban sewage sludge, sustainability, and transition for Eco-City: multi-criteria sustainability assessment of technologies based on best-worst method. Technol. Forecast. Soc. Chang. **116**, 29–39 (2017)

J. Ren, X. Ren, L. Dong, A. Manzardo, C. He, M. Pan, Multiactor multicriteria decision making for life cycle sustainability assessment under uncertainties. AIChE J. **64**(6), 2103–2112 (2018)

J. Rezaei, Best-worst multi-criteria decision-making method. Omega **53**, 49–57 (2015)

J. Rezaei, T. Nispeling, J. Sarkis, L. Tavasszy, A supplier selection life cycle approach integrating traditional and environmental criteria using the best worst method. J. Clean. Prod. **135**, 577–588 (2016)

J. Rezaei, J. Wang, L. Tavasszy, Linking supplier development to supplier segmentation using best worst method. Expert Syst. Appl. **42**(23), 9152–9164 (2015)

J. Rezaei, Best-worst multi-criteria decision-making method: some properties and a linear model. Omega **64**, 126–130 (2016)

T.L. Saaty, Decision making with the analytic hierarchy process. Int. J. Serv. Sci. **1**(1), 83–98 (2008)

S. Sahoo, A. Dhar, A. Kar, Environmental vulnerability assessment using grey analytic hierarchy process based model. Environ. Impact Assess. Rev. **56**, 145–154 (2016)

A. Sanayei, S.F. Mousavi, A. Yazdankhah, Group decision making process for supplier selection with VIKOR under fuzzy environment. Expert Syst. Appl. **37**(1), 24–30 (2010)

H.S. Shih, H.J. Shyur, E.S. Lee, An extension of TOPSIS for group decision making. Math. Comput. Model. **45**(7–8), 801–813 (2007)

R.K. Singh, S.K. Garg, S.G. Deshmukh, Interpretive structural modelling of factors for improving competitiveness of SMEs. Int. J. Prod. Qual. Manage. **2**(4), 423–440 (2007)

J. Thakkar, S.G. Deshmukh, A.D. Gupta, R. Shankar, Selection of third-party logistics (3PL): a hybrid approach using interpretive structural modeling (ISM) and analytic network process (ANP). Supply Chain Forum: Int. J. **6**(1), 32–46 (2005)

G. van de Kaa, L. Kamp, J. Rezaei, Selection of biomass thermochemical conversion technology in the Netherlands: a best worst method approach. J. Clean. Prod. **166**, 32–39 (2017)

C. Vasanthakumar, S. Vinodh, K. Ramesh, Application of interpretive structural modelling for analysis of factors influencing lean remanufacturing practices. Int. J. Prod. Res. **54**(24), 7439–7452 (2016)

G.H. Tzeng, J.J. Huang, *Multiple Attribute Decision Making: Methods and Applications* (Chapman and Hall/CRC, 2011)

Z. Wang, J. Ren, M.E. Goodsite, G. Xu, Waste-to-energy, municipal solid waste treatment, and best available technology: comprehensive evaluation by an interval-valued fuzzy multi-criteria decision making method. J. Clean. Prod. **172**, 887–899 (2018)

W.W. Wu, Y.T. Lee, Developing global managers' competencies using the fuzzy DEMATEL method. Expert Syst. Appl. **32**(2), 499–507 (2007)

X. Xia, K. Govindan, Q. Zhu, Analyzing internal barriers for automotive parts remanufacturers in China using grey-DEMATEL approach. J. Clean. Prod. **87**, 811–825 (2015)

# Chapter 2
# Two-Stage Interval Best-Worst Method for Weighting: Prioritization of Influential Factors of Airport Competitiveness

**Abstract** There are various factors influencing airport competitiveness, but it is usually difficult for the decision-makers to clearly understand the roles of these influential factors of airport competitiveness. In order to address this, this study aims at developing a two-stage interval best-worst method for determining the relative importance based on the multiplicative constraint. A total of twenty influential factors in five dimensions including airport capacity, network connectivity, service quality, operations and management, and external environment were summarized, then, the developed weighting method was employed to prioritize these influential factors, and they were categorized into three level, namely, significantly group, moderately important group, and less important group. Some policy implications were also proposed for building competitive airports and for improving the competitiveness of airports.

## 2.1 Introduction

Airport logistics which is a modern logistics relying on air transportation and airports developed rapidly with the development of the world's economy and the progress of society and human demand (Chen and Peng 2010). However, aviation industry faces both opportunities and threats currently under the conditions of modern logistics. The current conditions of airports for freight transport cannot satisfy the requirements of modern logistics (Shen and Yu 2012). In order to increase the market share of air transport, various studies were studied for improving the quality service. For instance, Chen (2016) selected the airline service quality improvement criteria based on Decision Making Trail and Evaluation Laboratory (DEMATEL) and Analytic Network Process, and the interdependent relationships among these criteria were also considered. Correia et al. (2008) developed a model for determining the relationship of the global index for the evaluation of the level of service with the individual components such as check-in and departure lounges through regression analysis. Pandey (2016) evaluated the service quality of airports in Thailand by using the fuzzy multi criteria decision making method, and the fuzzy expert system was also implemented to find the improvement areas of airport service quality. Barros and Wanke (2015) used the

 
J. Ren, *Advanced Operations Management for Complex Systems Analysis*,
SpringerBriefs in Applied Sciences and Technology,
https://doi.org/10.1007/978-3-030-45418-0_2

two-stage technique for order performance by similarity to ideal solution (TOPSIS) and data envelopment analysis (DEA) to investigate the African airlines efficiency. Lupo (2015) presented a fuzzy extension of the ServPerf service conceptual model to evaluate the service quality of international airports in Sicily based on ELECTRE III method. Bezerra and Gomes (2016) fitted the measurement model for perceived airport service quality, and the results revealed that the six-factor model could provide a meaningful multi-item instrument for measuring passenger perception of airport service quality. However, service quality is only one issue of airport competitiveness which can significantly influence the market of air transport. Therefore, more and more studies focused on the competitiveness of airports.

Part (1997) employed the fuzzy linguistic approach to assess the competitiveness of eight major airports in Asia. Park (2003) analysed the competitive status of the major airports in the East Asia region from five aspect including service, demand, managerial, facility, and spatial qualities, and the multi-criteria evaluation method was employed to determine the finally level of the competitiveness of the airports. Yeh et al. (2011) employed the fuzzy multiattribute decision making method to analyse the airport performance, the data with respect to the quantitative and qualitative criteria were scored by crisp and fuzzy numbers, respectively; and 11 Asia-Pacific major international airports were studied. Cui et al. (2013a, b) investigated the dynamic formation mechanism of airport competitiveness by using Structure Equation Model and System Dynamic. Chung and Han (2013) evaluated the competitiveness of transshipment cargo of three major airports in Northeast Asia including Incheon, Narita, and Pudong. Yeo et al. (2013) combined fuzzy AHP and the fuzzy technique for order performance by similarity to ideal solution (FTOPSIS) method to evaluate the competitiveness of the aerotropolises in East Asia, and five major aerotropolises including Beijing, Incheon, Shanghai, Taoyuan, and Hong Kong were investigated. Chao and Yu (2013) developed a quantitative evaluation model for evaluating the air cargo competitiveness through Delphi method to obtain the expert opinions. Cui et al. (2017) explored the difference of the airport competitiveness formation mechanism, and 45 airports in China during the period 2010–2014 were investigated. Chen and Peng (2010) employed AHP to evaluate the competitiveness of the airport logistics, the airports of Shanghai Pudong, Hong Kong, Nanjing Lukou, Ningbo Lishe, and Hangzhou were evaluated. It is apparent that most of the studies focused on competitiveness evaluation or ranking of different airports. However, there is no research focusing on investigating the influential factors of airport competitiveness, thus, it is difficult for the decision-makers/stakeholders to understand how to improve the competitiveness of the airports. In order to fill this research gap, this study aims at developing a multi-criteria analysis method for analyzing the relative importance of the influential factors for helping the stakeholders of airports to adopt some measures and strategies for improving the airport competitiveness.

Multi-criteria analysis method has been widely used in aviation. For instance, Garg (2016) developed a hybrid decision-support model by combing Analytic Hierarchy Process (AHP) and FTOPSIS for the selection of strategic alliance partner. Hsu and Liou (2013) developed a novel multi-criteria decision making model (so-called "DANP") by combining DEMATEL and ANP to select the best outsourcing provider. Zietsman and Vanderschuren (2014) employed AHP to assess the potential multi-airport systems in the case of Cape Town. Based on literature reviews, we could find that AHP was widely used for determining the weights of the criteria. Accordingly, it can also be employed to rank the influential factors of airport competitiveness. However, the AHP method and various methods derived from AHP (i.e. fuzzy AHP, grey AHP, and ANP) have two weak points: (i) the difficulty for guaranteeing the consistency of the comparison matrix for determining the weights of the criteria/factors; (ii) the difficulty of using the crisp numbers to express the preferences and opinions of the decision-making due to the vagueness, ambiguity and subjectivity existing in human's judgments. Rezaei (2015, 2016) developed a power weighting method-the best-worst method for determining the weights of the criteria/factors, and the best-worst method have two significant advantages, one is less times of computations compared with the AHP method and various methods derived from AHP, and another is the convenience for judgment consistency. Therefore, the best-worst method has been widely used recently. This study developed a two-stage interval best-worst method based on the work of Rezaei (2015, 2016) which allows the users to use interval numbers to compare the relative preference of one criterion over another to determine the weights of the criteria, and the developed method also have two advantages:

i. The interval numbers were used instead of the crisp numbers, and the characteristics of vagueness and ambiguity existing in human's judgments can be successfully solved;
ii. The first stage of the two-stage interval best-worst method is to minimize the inconsistency degree of the judgments, and the consistency can be guaranteed.

Besides the introduction, the remainder parts of this study has been organized as follows: Sect. 2.2 firstly presented the two-stage interval best-worst method; the influential factors of airport competitiveness were subsequently summarized in Sect. 2.3; the developed method was then applied on prioritizing the influential factors of airport competitiveness, and some policy implications were also proposed based on the obtained results in Sect. 2.4; finally, this study has been concluded in Sect. 2.5.

## 2.2 Two-Stage Interval Best-Worst Method

The traditional best-worst method was firstly introduced in Sect. 2.2.1; the interval number was introduced in Sect. 2.2.2; the developed two-stage interval best-worst method was presented in Sect. 2.2.3.

### 2.2.1 Traditional Best-Worst Method

The best-worst method developed by Rezaei (2015, 2016) consists of four steps, including determining the best and the worst criteria among all the evaluation criteria, determining the BO (best to others) and the OW (others to the worst) vector, determining the optimal weights of the criteria, and consistency check.

**Step 1**: Determining the best (e.g. most desirable, most preferable, and most important) and the worst (e.g. least desirable, least preferable, least important) criteria among all the evaluation criteria, denotes by $C_B$ and $C_W$, respectively.
**Step 2**: Comparing the best criterion with all the other criteria and all the other criteria with the worst criterion to determine the relative preferences by using the nine-scale system (the numbers from 1 to 9 and their reciprocals). Accordingly, both the Best-to-Others (BO) vector and the Others-to-Worst (OW) vector can be determined, as presented in Eq. (2.1) and Eq. (2.2), respectively.

$$BO = \left[ m_{B1}\ m_{B2}\ \ldots\ m_{Bn} \right] \tag{2.1}$$

$$OW = \left[ m_{1W}\ m_{2W}\ \ldots\ m_{nW} \right] \tag{2.2}$$

where $m_{Bj}(j = 1, 2, \ldots, n)$ and $m_{jW}(j = 1, 2, \ldots, n)$ represent the relative preference of the best criterion over the $j$-th criterion and that of the $j$-th criterion over the worst criterion, and n is the total number of the criteria.

It is apparent that when $j = B$, then $m_{Bj} = 1$, and when $j = W$, then $m_{jW} = 1$.

**Step 3**: Determining the optimal weights of the criteria.

The judgments in step 2 can be interpreted into Eqs. (2.3)–(2.4).

$$\frac{\omega_B}{\omega_j} = m_{Bj}(j = 1, 2, \ldots, n) \tag{2.3}$$

$$\frac{\omega_j}{\omega_W} = m_{jW}(j = 1, 2, \ldots, n) \tag{2.4}$$

where $\omega_j$ represents the weight of the $i$-th criterion, and $\omega_B$ and $\omega_W$ are the weights of the best and the worst criteria, respectively.

Equations (2.3)–(2.4) hold simultaneously only for absolutely consistent judgments, while they do not hold for inconsistent judgments. The programming for minimizing the maximum absolute difference $\left|\frac{\omega_B}{\omega_j} - m_{Bj}\right|$ and $\left|\frac{\omega_j}{\omega_W} - m_{jW}\right|$ for all j was developed for obtaining the weights of the criteria.

$$\min \max_j \left\{ \left| \frac{\omega_B}{\omega_j} - m_{Bj} \right|, \left| \frac{\omega_j}{\omega_W} - m_{jW} \right| \right\}$$

$$
\begin{aligned}
& s.t. \\
& \sum_{j=1}^{n} \omega_j = 1 \\
& \omega_j \geq 0, \quad j = 1, 2, \ldots, n
\end{aligned} \tag{2.5}
$$

Programming (2.5) can be rewritten as:

$$
\begin{aligned}
& \min \xi \\
& s.t. \\
& \left| \frac{\omega_B}{\omega_j} - m_{Bj} \right| \leq \xi, \quad j = 1, 2, \ldots, n \\
& \left| \frac{\omega_j}{\omega_W} - m_{jW} \right| \leq \xi, \quad j = 1, 2, \ldots, n \\
& \sum_{j=1}^{n} \omega_j = 1 \\
& \omega_j \geq 0, \quad j = 1, 2, \ldots, n
\end{aligned} \tag{2.6}
$$

where $\omega_B$ represents the weight of the best criterion, $\omega_W$ represent the weight of the worst criterion, and $\omega_j$ denotes the weight of the $j$-th criterion.

**Step 4**: Consistency check.

The consistency ratio can be calculated for consistency check, as presented in Eq. (2.7),

$$
CR = \frac{\xi^*}{CI} \tag{2.7}
$$

where CR represents the consistency ratio, and CR represents the consistency index.

The value of CI which depends on $a_{BW}$ can be determined according to Table 2.1, and the value of consistency ratio varies with the interval $\begin{bmatrix} 0 & 1 \end{bmatrix}$, and it indicates the consistency degree, and the smaller the value, the more consistent the judgments are; in contrary, the greater the value, the more inconsistent the judgments are.

The best-worst method is a novel, powerful and convenient tool for weights determination with the two most significant advantages: one is fewer times of comparisons comparing with the AHP method and various methods derived from AHP. However,

**Table 2.1** Consistency index (CI) table (Rezaei 2015, 2016)

| $a_{BW}$ | 1 | 2 | 3 | 4 | 5 | 6 | 7 | 8 | 9 |
|---|---|---|---|---|---|---|---|---|---|
| Consistency index (max $\xi$) | 0.00 | 0.44 | 1.00 | 1.63 | 2.30 | 3.00 | 3.73 | 4.47 | 5.23 |

**Table 2.2** The nine scale system in Saaty method (Saaty 1978)

| Scale | Definition | Note |
|---|---|---|
| 1 | Equal importance | $i$ is equally important to $j$ |
| 3 | Moderate importance | $i$ is moderately important to $j$ |
| 5 | Essential importance | $i$ is essentially important to $j$ |
| 7 | Very strong importance | $i$ is very strongly important to $j$ |
| 9 | Absolute importance | $i$ is very absolutely important to $j$ |
| 2, 4, 6, 8 | Intermediate value | The relative importance of $i$ to $j$ is between the two adjacent judgments |

it relies on using the crisp numbers (the numbers from 1 to 9 and their reciprocals) to establish the BO and OW vectors; however, it is usually difficult for the users to use crisp numbers to express their preferences/opinions between each pair of criteria, because there are various types of vagueness, ambiguity and subjectivity existing in human judgments. For instance, the users hold the view that the relative preference of a criterion over another cannot be directly expressed by a single linguistic term presented in the Saaty method (see Table 2.2), e.g. the relative preference is between "moderate importance" (corresponding to 3) and "essential importance" (corresponding to 5), thus, the interval $\left[3\ 5\right]$ should be used to describe the relative preference. Therefore, the interval numbers are more convenient for the users to establish the BO and OW vectors. The interval numbers were introduction in Sect. 2.2.

### 2.2.2 *Interval Numbers*

Let $x = \left[x^- \ x^+\right] = \left\{x \middle| x^- \le x \le x^+, x^- \le x^+, x^-, x^+ \in \mathbb{R}\right\}$. Then, $x = \left[x^- \ x^+\right]$ is called an interval number. When $0 \le x^- \le x^+$, $x = \left[x^- \ x^+\right]$ is a positive interval number.

**Definition 1** Midpoint of Interval number (Alefeld and Herzberge 2015)

The midpoint of the interval number $x = \left[x^- \ x^+\right]$ can be determined by Eq. (2.8).

$$m(x) = \frac{x^- + x^+}{2} \tag{2.8}$$

where $m(x)$ represents the midpoint of $x = \left[x^- \ x^+\right]$.

If $a = \left[a^- \ a^+\right]$ and $b = \left[b^- \ b^+\right]$ are two positive interval numbers, is a positive crisp number, the arithmetic operations between them are presented as follows.

**Definition 2** Arithmetic operations (Alefeld and Herzberge 2015; Xu 2008)

$$a + b = \left[a^- \ a^+\right] + \left[b^- \ b^+\right] = \left[a^- + b^-, a^+ + b^+\right] \tag{2.9}$$

$$a - b = \left[a^- \; a^+\right] - \left[b^- \; b^+\right] = \left[a^- - b^+, a^+ - b^-\right] \tag{2.10}$$

$$a \times b = \left[a^- \; a^+\right] \times \left[b^- \; b^+\right] = \left[a^- b^-, a^+ b^+\right] \tag{2.11}$$

$$a/b = \left[a^- \; a^+\right] / \left[b^- \; b^+\right] = \left[\frac{a^-}{b^+}, \frac{a^+}{b^-}\right] \tag{2.12}$$

$$\lambda a = \lambda\left[a^- \; a^+\right] = \left[\lambda a^- \; \lambda a^+\right] \tag{2.13}$$

$$a^\lambda = \left[a^- \; a^+\right]^\lambda = \left[\left(a^-\right)^\lambda, \left(a^+\right)^\lambda\right] \tag{2.14}$$

### 2.2.3 *Two-Stage Interval Best-Worst Method*

A two-stage interval best-worst method in which the users are allowed to used interval numbers to establish the BO and OW vectors was developed in this study, the first stage is to minimize the inconsistency degree, and the second stage is to determine the optimal weights of the criteria. It consists of four steps, they were specified as follows based on the work of Rezaei (2015, 2016) and Wang et al. (2005):

**Step 1**: Determining the best and the worst criteria. This step is the same with that in the traditional BW method, the users will firstly determine the best and the worst criteria among all the criteria according to their priorities.
**Step 2**: Determining the interval BO and OW vectors. The users will employ the interval numbers composed by the crisp numbers presented in Table 2.2 to establish the BO and OW vectors, as presented in Eq. (2.15) and Eq. (2.16), respectively.

$$BO = \left[\left[l_{B1} \; u_{B1}\right] \left[l_{B2} \; u_{B2}\right] \ldots \left[l_{Bn} \; u_{Bn}\right]\right] \tag{2.15}$$

$$OW = \left[\left[l_{1W} \; u_{1W}\right] \left[l_{2W} \; u_{2W}\right] \ldots \left[l_{nW} \; u_{nW}\right]\right] \tag{2.16}$$

Note that when $j = B$, then $\left[l_{Bj} \; u_{Bj}\right] = \left[1 \; 1\right]$, and when $j = W$, then $\left[l_{jW} \; u_{jW}\right] = \left[1 \; 1\right]$.

**Step 3**: Minimizing the inconsistency degree.

The judgments in step 2 can also be interpreted into Eqs. (2.17)–(2.18).

$$l_{Bj} \leq \frac{\omega_B}{\omega_j} \leq u_{Bj} (j = 1, 2, \ldots, n) \tag{2.17}$$

$$l_{jW} \leq \frac{\omega_j}{\omega_W} \leq u_{jW}(j = 1, 2, \ldots, n) \tag{2.18}$$

Inequalities (2.17) and (2.18) can be rewritten as

$$\ln l_{Bj} \leq \ln \omega_B - \ln \omega_j \leq \ln u_{Bj}, \quad j = 1, 2, \ldots, n \tag{2.19}$$

$$\ln l_{jW} \leq \ln \omega_j - \ln \omega_W \leq \ln u_{jW} \quad j = 1, 2, \ldots, n \tag{2.20}$$

Inequalities (2.19) and (2.20) hold only for consistent judgment but not for the inconsistent judgments. The derivation variables $p_{Bj}(j = 1, 2, \ldots, n)$, $q_{Bj}(j = 1, 2, \ldots, n)$, $e_{jW}(j = 1, 2, \ldots, n)$, and $f_{jW}(j = 1, 2, \ldots, n)$ were introduced to make these inequalities satisfy both consistent and inconsistent judgments, as presented in (2.21) and (2.22),

$$\ln l_{Bj} - p_{Bj} \leq \ln \omega_B - \ln \omega_j \leq \ln u_{Bj} + q_{Bj}, \quad j = 1, 2, \ldots, n \tag{2.21}$$

$$\ln l_{jW} - e_{jW} \leq \ln \omega_j - \ln \omega_W \leq \ln u_{jW} + f_{jW}, \quad j = 1, 2, \ldots, n \tag{2.22}$$

$p_{Bj}$ and $q_{Bj}$ are two nonnegative real numbers, but only one of them can be positive, namely, $p_{Bj}q_{Bj} = 0$. Similarly, $e_{jW}$ and $f_{jW}$ are also two nonnegative real numbers, but only one of them can be positive, namely, $e_{jW} f_{jW} = 0$.

The multiplicative constraint which was widely used as presented in Eq. (2.23) was used as the constraint for determining the weights of the criteria.

$$\prod_{j=1}^{n} \omega_j = 1 \tag{2.23}$$

The derivation variables should be kept as small as possible for a better consistency. Accordingly, the following programming can be obtained to determine the weights of the criteria with the consideration of the multiplicative constraint.

$$\begin{aligned}
&\min J = \sum \left(p_{Bj} + q_{Bj} + e_{jW} + f_{jW}\right) \\
&\ln \omega_B - \ln \omega_j + p_{Bj} \geq \ln l_{Bj}, \quad j = 1, 2, \ldots, n, j \neq B \\
&\ln \omega_B - \ln \omega_j - q_{Bj} \leq \ln u_{Bj}, \quad j = 1, 2, \ldots, n, j \neq B \\
&\ln \omega_j - \ln \omega_W + e_{jW} \geq \ln l_{jW}, \quad j = 1, 2, \ldots, n, j \neq B, W \\
&\ln \omega_j - \ln \omega_W - f_{jW} \leq \ln u_{jW}, \quad j = 1, 2, \ldots, n, j \neq B, W \\
&\sum_{j=1}^{n} \ln \omega_j = 0 \\
&p_{Bj}q_{Bj} = 0, \quad j = 1, 2, \ldots n, j \neq B \\
&e_{jW} f_{jW} = 0, \quad j = 1, 2, \ldots, n, j \neq B, W
\end{aligned}$$

$$p_{Bj}, q_{Bj} \geq 0, \quad j = 1, 2, \ldots, n, j \neq B$$
$$e_{jW}, f_{jW} \geq 0, \quad j = 1, 2, \ldots, n, j \neq B, W \tag{2.24}$$

The logarithms were used to substitute the original judgments, because $\ln \omega_j$ is nonnegative when $\omega_j$ is greater than or equal to 1, while $\ln \omega_j$ is negative when $\omega_j$ is smaller than 1, the following nonnegative variables were introduced in (2.24).

$$a_i = \frac{\ln \omega_j + |\ln \omega_j|}{2} \tag{2.25}$$

$$b_i = \frac{-\ln \omega_j + |\ln \omega_j|}{2} \tag{2.26}$$

Accordingly, $\ln \omega_j$ can be rewritten as:

$$\ln \omega_j = a_j - b_j \quad j = 1, 2, \ldots, n \tag{2.27}$$

where $a_j b_j = 0$.

The programming (2.24) can be further rewritten as (2.28), and the objective function can be recognized as the total inconsistency degree.

$$\begin{aligned}
&\min J = \sum \left(p_{Bj} + q_{Bj} + e_{jW} + f_{jW}\right) \\
&a_B - b_B - a_j + b_j + p_{Bj} \geq \ln l_{Bj}, \quad j = 1, 2, \ldots, n, j \neq B \\
&a_B - b_B - a_j + b_j - q_{Bj} \leq \ln u_{Bj}, \quad j = 1, 2, \ldots, n, j \neq B \\
&a_j - b_j - a_W + b_W + e_{jW} \geq \ln l_{jW}, \quad j = 1, 2, \ldots, n, j \neq B, W \\
&a_j - b_j - a_W + b_W - f_{jW} \leq \ln u_{jW}, \quad j = 1, 2, \ldots, n, j \neq B, W \\
&\sum_{j-1}^{n} \left(a_j - b_j\right) = 0 \\
&p_{Bj} q_{Bj} = 0, \quad j = 1, 2, \ldots, n, j \neq B \\
&e_{jW} f_{jW} = 0, \quad j = 1, 2, \ldots, n, j \neq B, W \\
&a_j b_j = 0, \quad j = 1, 2, \ldots, n \\
&p_{Bj}, q_{Bj} \geq 0, \quad j = 1, 2, \ldots, n, j \neq B \\
&e_{jW}, f_{jW} \geq 0, \quad j = 1, 2, \ldots, n, j \neq B, W \\
&a_j, b_j \geq 0, \quad j = 1, 2, \ldots, n
\end{aligned} \tag{2.28}$$

After solving programming (2.24), the minimum value of $J$ can be obtained, denotes by $J^*$, and it means that the inconsistency is the least under this condition. If and only if $J^* = 0$ the judgments are consistent.

**Step 4**: Determining the optimal interval weights of the criteria.

Keeping the inconsistency degree the minimum, the following programming models were established to determine the interval of each weight.

$$\begin{aligned}
&\min/\max \ln \omega_j = a_i - b_i \\
&a_B - b_B - a_j + b_j + p_{Bj} \geq \ln l_{Bj}, \quad j = 1, 2, \ldots, n, j \neq B \\
&a_B - b_B - a_j + b_j - q_{Bj} \leq \ln u_{Bj}, \quad j = 1, 2, \ldots, n, j \neq B \\
&a_j - b_j - a_W + b_W + e_{jW} \geq \ln l_{jW}, \quad j = 1, 2, \ldots, n, j \neq B, W \\
&a_j - b_j - a_W + b_W - f_{jW} \leq \ln u_{jW}, \quad j = 1, 2, \ldots, n, j \neq B, W \\
&\sum_{j=1}^{n} \left(a_j - b_j\right) = 0 \\
&\sum_{j=1,,j \neq W}^{n} \left(p_{Bj} + q_{Bj} + e_{jW} + f_{jW}\right) = J^* \\
&p_{Bj} q_{Bj} = 0, \quad j = 1, 2, \ldots, n, j \neq B \\
&e_{jW} f_{jW} = 0, \quad j = 1, 2, \ldots, n, j \neq B, W \\
&a_j b_j = 0, \quad j = 1, 2, \ldots, n \\
&p_{Bj}, q_{Bj} \geq 0, \quad j = 1, 2, \ldots, n, j \neq B \\
&e_{jW}, f_{jW} \geq 0, \quad j = 1, 2, \ldots, n, j \neq B, W \\
&a_j, b_j \geq 0, \quad j = 1, 2, \ldots, n
\end{aligned} \tag{2.29}$$

After solving programming (2.29), the lower and upper bounds of $\ln \omega_j$ can be determined, denoted by $\ln \omega_j^L$ and $\ln \omega_j^U$, respectively, then, the interval weight of each criterion can be determined, as presented in Eq. (2.30).

$$\left[\omega_j^L \ \omega_j^U\right] = \left[\exp\left(\ln \omega_j^L\right) \ \exp\left(\ln \omega_j^U\right)\right] \tag{2.30}$$

It is apparent that if $J^* = 0$, programming (2.29) can be simplified as:

$$\begin{aligned}
&\min/\max \ln \omega_j = a_i - b_i \\
&a_B - b_B - a_j + b_j \geq \ln l_{Bj}, \quad j = 1, 2, \ldots, n, j \neq B \\
&a_B - b_B - a_j + b_j \leq \ln u_{Bj}, \quad j = 1, 2, \ldots, n, j \neq B \\
&a_j - b_j - a_W + b_W \geq \ln l_{jW}, \quad j = 1, 2, \ldots, n, j \neq B, W \\
&a_j - b_j - a_W + b_W \leq \ln u_{jW}, \quad j = 1, 2, \ldots, n, j \neq B, W \\
&\sum_{j=1}^{n} \left(a_j - b_j\right) = 0 \\
&a_j b_j = 0, \quad j = 1, 2, \ldots, n \\
&a_j, b_j \geq 0, \quad j = 1, 2, \ldots, n
\end{aligned} \tag{2.31}$$

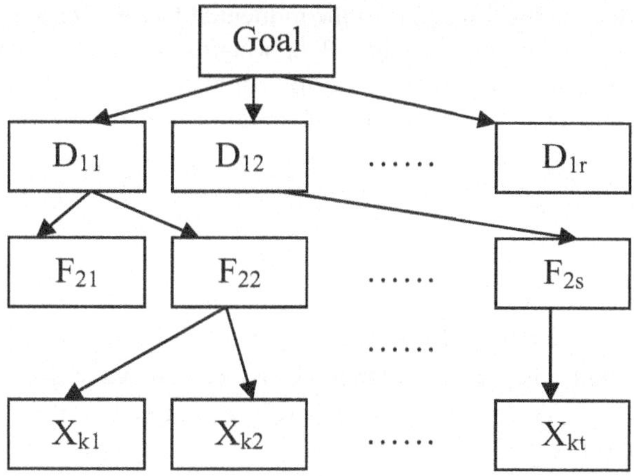


**Fig. 2.1** Hierarchical structure (An et al. 2017)

As for the problem with multiple hierarchies (as presented in Fig. 2.1), the global weights of the influential factors can be determined in the following way.

Suppose there are total T dimensions (t = 1, 2, …, T) in the *L*-th hierarchy, denotes by $D_t$ (t = 1, 2, …, T), and there are a total of K influential factors (k = 1, 2, …, K) in the *t*-th dimension belonging to the (*L* + 1)-th hierarchy, denotes by $F_{tk}$(k = 1, 2, …, K), Eqs. (2.32)–(2.33) could be obtained according to the multiplicative principle (Wang et al. 2005).

$$\frac{\omega_1}{\omega_{t1}} = \frac{\omega_2}{\omega_{t2}} = \cdots = \frac{\omega_k}{\omega_{tk}} = \cdots = \frac{\omega_K}{\omega_{tK}} \tag{2.32}$$

$$\prod_{k=1}^{K} \omega_k = \omega_t \tag{2.33}$$

According to Eqs. (2.31) (2.32), the formula for determining the global eight of the *k*-th factor to the *t*-th dimension belonging to the *L*-th hierarchy can be obtained, as presented in Eq. (2.34).

$$\omega_k = \left[ \frac{\omega_t \times (\omega_{tk})^K}{\prod_{r=1}^{K} \omega_{tr}} \right]^{1/K} \tag{2.34}$$

where $\omega_k$ represents the global weight of the *k*-th factor to the *t*-th dimension belonging to the *L*-th hierarchy, $\omega_t$ represents the weight of the *t*-th dimension belonging to the (*L* + 1)-th hierarchy, and $\omega_{tk}$ represents the local weight of the *k*-th factor in the *t*-th dimension.

However, Eq. (2.33) can only be used to determine the global weights of the influential factors when all the weights including both the weights of the dimensions and the local weights of the influential factors in each dimension are crisp numbers. As for the condition that all the weights including both the weights of the dimensions and the local weights of the influential factors in each dimension are interval numbers, the global weights of the influential factors can be determined by Eq. (2.35).

$$\left[\omega_k^- \ \omega_k^+\right] = \left[\left[\frac{\omega_t^- \times \left(\omega_{tk}^-\right)^K}{\prod_{r=1}^{K} \omega_{tr}^+}\right]^{1/K} \left[\frac{\omega_t^+ \times \left(\omega_{tk}^+\right)^K}{\prod_{r=1}^{K} \omega_{tr}^-}\right]^{1/K}\right] \tag{2.35}$$

As for two global interval weights $\left[\omega_k^- \ \omega_k^+\right]$ and $\left[\omega_s^- \ \omega_s^+\right]$ which represent the relative importance of the $k$-th and the $s$-th factor, the probability degree of $\left[\omega_k^- \ \omega_k^+\right]$ be greater than $\left[\omega_s^- \ \omega_s^+\right]$ can be determined by Eq. (2.36) according to Zhou et al. (2012).

$$\begin{aligned} p_{ks} &= P\left(\left[\omega_k^- \ \omega_k^+\right] \geq \left[\omega_s^- \ \omega_s^+\right]\right) \\ &= \max\left\{1 - \frac{1}{2}\max\left[\frac{\omega_s^- + \omega_s^+ - \left(\omega_k^- + \omega_k^+\right)}{\omega_s^+ - \omega_s^- + \omega_k^+ - \omega_k^-} + 1, 0\right], 0\right\} \end{aligned} \tag{2.36}$$

where $p_{ks} = P\left(\left[\omega_k^- \ \omega_k^+\right] \geq \left[\omega_s^- \ \omega_s^+\right]\right)$ represents the probability degree of $\left[\omega_k^- \ \omega_k^+\right]$ be greater than $\left[\omega_s^- \ \omega_s^+\right]$.

Suppose that there are total n factors, the probability degree matrix can be determined by comparing the global interval weights of each pair of factors, as presented in Eq. (2.37).

$$P = \begin{bmatrix} p_{11} & p_{12} & \cdots & p_{1n} \\ p_{21} & p_{22} & \cdots & p_{2n} \\ \vdots & \vdots & \ddots & \vdots \\ p_{n1} & p_{n2} & \cdots & p_{nn} \end{bmatrix} \tag{2.37}$$

where $p_{ks}(k = 1, 2, \ldots, n; s = 1, 2, \ldots, n)$ represents the probability degree of the interval weight of the $k$-th factor be greater than that of the $s$-th factor.

According to the work of Xu and Da (2003), these interval weights can be ranked according to their relative priorities (RP), as presented in Eq. (2.38).

$$RP_k = \frac{\sum_{s=1}^{n} p_{ks} + \frac{n}{2} - 1}{n(n-1)} \tag{2.38}$$

where $RP_k$ represents the relative priority of the k-th factor.

## 2.3 Influential Factors of Airport Competitiveness

In order to summarize the influential factors of airport competitiveness, a comprehensive literature review through searching the key words such as "airport", "competitiveness", and "service quality" in Google Scholar and CNKI (China National Knowledge Infrastructure) was carried out, and then a focus group meeting was held to screen the influential factors of airport competitiveness. The influential factors of airport competitiveness were selected based on the following three principles:

1. Representative principle: all the influential factors selected should be representative to avoid repetitions and redundancy. In others, the selection of each influential factor should avoid the overlaps in concepts and definitions with some other influential factors;
2. Systematic principle: the selected influential factors should be enough to represent each dimension of the airport competitiveness;
3. Object-oriented principle: the selection of the influential factors of airport competitiveness should consider the preferences and willingness of the stakeholders.

According to the above mentioned three principles, a total of 20 influential factors in 5 dimensions were determined to measure the airport competitiveness in China (see Table 2.3), the five dimensions are airport capacity ($D_1$), network connectivity ($D_2$), service quality ($D_3$), operations and management ($D_4$), and external environment ($D_5$) (Peng and Li 2011). The influential factors in each dimension were summarized as follows.

1. Airport capacity ($D_1$): this dimension is to reflect the airport infrastructure and transport capacity, and four influential factors were considered in this dimension, and they are airport infrastructure, passenger throughput, cargo throughput, and number of takeoffs and landings in the peak hour.

   - Airport infrastructure ($D_{11}$): the airport infrastructure mainly refers to the airport runway number, boarding gate number, area of the terminal building, and number of gate positions(Peng and Li 2011).
   - Passenger throughput ($D_{12}$): the total passenger throughput per year can represent the operation scale of an airport in passenger transport industry, and it is also a measure of the competitiveness performance on market (Su 2011).
   - Cargo throughput ($D_{13}$): the total cargo throughput per year can represent the operation scale of an airport in freight transport industry, and it is also a measure of the competitiveness performance on market (Su 2011).
   - Number of takeoffs and landings in the peak hour ($D_{14}$): the number of takeoffs and landings in the peak hour can reflect the infrastructure level of the airports, the transport capacity, and the ATC efficiency per airspace (Su et al. 2010; Zhao 2016).

**Table 2.3** The influential factors of airport competitiveness in China

| Dimensions | | References |
|---|---|---|
| Airport capacity ($D_1$) | Airport infrastructure ($D_{11}$) | Peng and Li (2011) |
| | Passenger throughput ($D_{12}$) | Su (2011) |
| | Cargo throughput ($D_{13}$) | Su (2011) |
| | Number of takeoffs and landings in the peak hour ($D_{14}$) | Su et al. (2010), Zhao (2016) |
| Network connectivity ($D_2$) | Number of navigable cities ($D_{21}$) | Peng and Li (2011) |
| | Percentage of transit passenger ($D_{22}$) | Dong et al. (2007) |
| | Flight density ($D_{23}$) | Dong et al. (2007), Peng and Li (2011) |
| | Number of airline companies ($D_{24}$) | Liang et al. (2016) |
| Service quality ($D_3$) | Airport security and safety ($D_{31}$) | Peng and Li (2011) |
| | On-time performance of flights ($D_{32}$) | Dong et al. (2007) |
| | Service satisfaction rate ($D_{33}$) | Su et al. (2010) |
| | Logistic service ($D_{34}$) | Su (2011) |
| | Percentage of ground crew with college degree ($D_{35}$) | Cui et al. (2013) |
| Operations and management ($D_4$) | Time for flight transfer ($D_{41}$) | Peng and Li (2011) |
| | Time for check-in, security check and baggage claim ($D_{42}$) | Xie et al. (2015) |
| | The income of main business per capita ($D_{43}$) | Dong et al. (2007) |
| External environment ($D_5$) | Geographical conditions ($D_{51}$) | Su et al. (2010) |
| | Governmental support ($D_{52}$) | Peng and Li (2011) |
| | GDP of the cities within the service radius of the airport ($D_{53}$) | Dong et al. (2007) |
| | Population of the cities within the service radius of the airport ($D_{54}$) | Cui et al. (2013) |

2. Network connectivity ($D_2$): this dimension is to reflect the ability of the airports to satisfy the requirements of the customers and the airline network status for passenger and freight transport.

   - Number of navigable cities ($D_{21}$): it represents the number of the cities which has regular flights between the airport and them, and it can reflect the network connectivity of the airlines (Peng and Li 2011).
   - Percentage of transit passenger ($D_{22}$): the percentage of transit passenger can be determined by calculating the ratio between the transit passengers with passenger throughput, and it can reflect the roles of the airports in the network connectivity (Dong et al. 2007).

- Flight density ($D_{23}$): this criterion represents the total flights in a period, and it can reflect the network connectivity of the airports from the perspective of flights numbers (Dong et al. 2007; Peng and Li 2011).
- Number of airline companies ($D_{24}$): the number of airline companies can reflect the operating scale of the airports (Liang et al. 2016).

3. Service quality ($D_3$)

- Airport security and safety ($D_{31}$): this criterion is to measure the airport management level for security and safety, including objective system of safety management, organization structure of safety management, regulation system for security and safety, education and training on security and safety, and equipment for security and safety (Peng and Li 2011).
- On-time performance of flights ($D_{32}$): this criterion can directly reflect the service quality and efficiency of airports (Dong et al. 2007).
- Service satisfaction rate ($D_{33}$): the service satisfaction rate can be determined by calculating the ratio between the satisfied customers, owners of cargo, and airline companies with the total serviced people (Su et al. 2010).
- Logistic service ($D_{34}$): this criterion can be measured by the informatization level and the waiting area per capita in the peak hour (Su 2011).
- Percentage of ground crew with college degree ($D_{35}$): this refers to the ratio of the number of the ground crew with college degree with the number of the total ground crew, and this criterion can reflect the overall education and training level of the ground crew (Cui et al. 2013a, b).

4. Operations and management ($D_4$): this dimension is to reflect the operations and management level of the airports.

- Time for flight transfer ($D_{41}$): this refers to the time of the flights from arrival to departure, and it can reflect the convenience level of the airports (Peng and Li 2011).
- Time for check-in, security check and baggage claim ($D_{42}$): this criterion can effectively reflect the convenience of the airports for the passengers (Xie et al. 2015).
- The income of main business per capita ($D_{43}$): this criterion can reflect the operating benefits, the use status of asset and the future financial situation of the airports (Dong et al. 2007).

5. External environment ($D_5$): this dimension is to measure the external environment and factors that influence the competitiveness of the airports.

- Geographical conditions ($D_{51}$): the locations of the airports with different geographical conditions which can reflect the suitability of the airports for developing transport industry can significantly influence the roles of the airports as traffic hub (Su et al. 2010).

- Governmental support ($D_{52}$): the governmental support represents the policy and regulation support from the government in tax and land use (Peng and Li 2011).
- GDP of the cities within the service radius of the airport ($D_{53}$): the GDP of the cities within the service radius of the airport can reflect the economy development status of the city, and can further reflect the demand for passenger transport and cargo transport (Dong et al. 2007).
- Population of the cities within the service radius of the airport ($D_{54}$): the total population of the cities within the service radius of the airport can reflect the potential of passenger who will use the corresponding airport (Cui et al. 2013a, b).

## 2.4 Results and Policy Implications

The interval multiplicative best-worst method presented in Sect. 2.2.3 was employed to prioritize these influential factors by determining their relative importance. Taking the determining the interval weights of the five dimensions as an example, these four steps were illustrated as follows:

**Step 1**: Determining the most important and the least important dimensions among these five dimensions. External environment ($D_5$) and operations and management ($D_4$) were recognized as the most important and the least important dimensions in the focus group meeting.
**Step 2**: Determining the interval BO and OW vectors. The interval BO and OW vectors were determined, as presented in Table 2.4.
**Step 3**: Minimizing the inconsistency degree. The programming for minimizing the inconsistency degree of the judgments was established according to (2.28), as presented in (2.39). After substituting the values of the elements in the BO and OW vectors, this programming can be solved, and then the minimum inconsistency degree can be determined, namely, $J^* = 0$.

**Table 2.4** The interval BO and OW vectors for determining the interval weights of the five dimensions of airport competitiveness

| | Airport capacity ($D_1$) | Network connectivity ($D_2$) | Service quality ($D_3$) | Operations and management ($D_4$) | External environment ($D_5$) |
|---|---|---|---|---|---|
| | | | | Least important | Most important |
| BO | $[2\ 4]$ | $[1\ 2]$ | $[2\ 3]$ | $[4\ 6]$ | $[1\ 1]$ |
| OW | $[1\ 3]$ | $[3\ 5]$ | $[2\ 4]$ | $[1\ 1]$ | $[4\ 6]$ |

$$
\begin{aligned}
&\min J = (p_{51} + q_{51} + p_{52} + q_{52} + p_{53} + q_{53} \\
&\qquad + p_{54} + q_{54} + e_{14} + f_{14} + e_{24} + f_{24} + e_{34} + f_{34}) \\
&a_5 - b_5 - a_1 + b_1 + p_{51} \geq \ln l_{51} \\
&a_5 - b_5 - a_1 + b_1 - q_{51} \leq \ln u_{51} \\
&a_5 - b_5 - a_2 + b_2 + p_{52} \geq \ln l_{52} \\
&a_5 - b_5 - a_2 + b_2 - q_{52} \leq \ln u_{52} \\
&a_5 - b_5 - a_3 + b_3 + p_{53} \geq \ln l_{53} \\
&a_5 - b_5 - a_3 + b_3 - q_{53} \leq \ln u_{53} \\
&a_5 - b_5 - a_4 + b_4 + p_{53} \geq \ln l_{54} \\
&a_5 - b_5 - a_4 + b_4 - q_{53} \leq \ln u_{54} \\
&a_1 - b_1 - a_4 + b_4 + e_{14} \geq \ln l_{14} \\
&a_1 - b_1 - a_4 + b_4 - f_{14} \leq \ln u_{14} \\
&a_2 - b_2 - a_4 + b_4 + e_{24} \geq \ln l_{24} \\
&a_2 - b_2 - a_4 + b_4 - f_{24} \leq \ln u_{24} \\
&a_3 - b_3 - a_4 + b_4 + e_{14} \geq \ln l_{34} \\
&a_3 - b_3 - a_4 + b_4 - f_{34} \leq \ln u_{34} \\
&\sum_{j=1}^{5} (a_j - b_j) = 0 \\
&p_{51}q_{51} = 0,\ p_{52}q_{52} = 0,\ p_{53}q_{53} = 0,\ p_{54}q_{54} = 0 \\
&e_{14}f_{14} = 0,\ e_{24}f_{24} = 0,\ e_{34}f_{34} = 0 \\
&a_1b_1 = 0,\ a_2b_2 = 0,\ a_3b_3 = 0,\ a_4b_4 = 0,\ a_5b_5 = 0 \\
&p_{51}, q_{51}, p_{52}, q_{52}, p_{53}, q_{53}, p_{54}, q_{54} \geq 0 \\
&e_{14}, f_{14}, e_{24}, f_{24}, e_{34}, f_{34} \geq 0 \\
&a_1, b_1, a_2, b_2, a_3, b_3, a_4, b_4, a_5, b_5 \geq 0
\end{aligned} \tag{2.39}
$$

**Step 4**: Determining the optimal interval weights of the five dimensions. As $J^* = 0$, programming (2.31) can be employed to determine the interval weights of the five dimensions. Taking the interval weight of the airport capacity ($D_1$) as an example, the programming (2.40) can be established according to programming (2.31).

$$
\begin{aligned}
&\min/\max \ln \omega_1 = (a_1 - b_1) \\
&a_5 - b_5 - a_1 + b_1 \geq \ln l_{51} \\
&a_5 - b_5 - a_1 + b_1 \leq \ln u_{51} \\
&a_5 - b_5 - a_2 + b_2 \geq \ln l_{52} \\
&a_5 - b_5 - a_2 + b_2 \leq \ln u_{52} \\
&a_5 - b_5 - a_3 + b_3 \geq \ln l_{53}
\end{aligned}
$$

$$
\begin{aligned}
& a_5 - b_5 - a_3 + b_3 \leq \ln u_{53} \\
& a_5 - b_5 - a_4 + b_4 \geq \ln l_{54} \\
& a_5 - b_5 - a_4 + b_4 \leq \ln u_{54} \\
& a_1 - b_1 - a_4 + b_4 \geq \ln l_{14} \\
& a_1 - b_1 - a_4 + b_4 \leq \ln u_{14} \\
& a_2 - b_2 - a_4 + b_4 \geq \ln l_{24} \\
& a_2 - b_2 - a_4 + b_4 \leq \ln u_{24} \\
& a_3 - b_3 - a_4 + b_4 \geq \ln l_{34} \\
& a_3 - b_3 - a_4 + b_4 \leq \ln u_{34} \\
& \sum_{j=1}^{5} (a_j - b_j) = 0 \\
& a_1 b_1 = 0, a_2 b_2 = 0, a_3 b_3 = 0, a_4 b_4 = 0, a_5 b_5 = 0 \\
& a_1, b_1, a_2, b_2, a_3, b_3, a_4, b_4, a_5, b_5 \geq 0
\end{aligned}
\tag{2.40}
$$

After solving programming (2.40), $\ln \omega_1^L$ and $\ln \omega_1^U$ can be determined, and they are −0.6931 and 0.1622, respectively. Then, the interval weight of the dimension 'airport capacity ($D_1$)' can be determined, as presented in Eq. (2.41).

$$\left[ \omega_j^L \; \omega_j^U \right] = \left[ \exp(-0.6931) \; \exp(0.1622) \right] = \left[ 0.5000 \; 1.1761 \right] \tag{2.41}$$

In a similar way, the weights of all the five dimensions can be determined, as presented in Table 2.5.

Similarly, the local weights of the factors in each dimension can also be determined, and the results were presented in Tables 2.6, 2.7, 2.8, 2.9 and 2.10.

After determining the interval weights of the five dimensions and the local interval weights of the influential factors in each dimension, the global interval weights of the influential factors can be determined by Eq. (2.35). Taking the global interval weight of airport infrastructure ($D_{11}$) in airport capacity ($D_1$) as an example, the interval weight of the dimension-airport capacity ($D_1$) is $\left[ 0.5000 \; 1.1761 \right]$, and the local interval weights of airport infrastructure ($D_{11}$), passenger throughput ($D_{12}$), cargo throughput ($D_{13}$), and number of takeoffs and landings in the peak hour ($D_{14}$) are $\left[ 2.2134 \; 2.4846 \right]$, $\left[ 1.0246 \; 1.1954 \right]$, $\left[ 1.0246 \; 1.1954 \right]$, and $\left[ 0.3286 \; 0.3689 \right]$, respectively. Then, Eq. (2.42) can be formulated to determine the global weights of airport infrastructure ($D_{11}$).

$$
\begin{aligned}
& \left[ \left[ \frac{0.5000 \times (2.2134)^4}{2.4846 \times 1.1954 \times 1.1954 \times 0.3689} \right]^{1/4} \left[ \frac{1.1761 \times (2.4846)^K}{2.2134 \times 1.0246 \times 1.0246 \times 0.3286} \right]^{1/4} \right] \\
& = \left[ 1.7398 \; 2.7680 \right]
\end{aligned}
\tag{2.42}
$$

**Table 2.5** The weights of the five dimensions of airport competitiveness

| Dimensions | Airport capacity ($D_1$) | Network connectivity ($D_2$) | Service quality ($D_3$) | Operations and management ($D_4$) | External environment ($D_5$) |
|---|---|---|---|---|---|
| Interval weights | $[0.5000\ 1.1761]$ | $[0.9221\ 2.1867]$ | $[0.7079\ 1.2457]$ | $[0.3264\ 0.5296]$ | $[1.7411\ 2.7019]$ |

**Table 2.6** The local weights of the factors in the dimension of airport capacity ($D_1$)

| | Airport infrastructure ($D_{11}$) | Passenger throughput ($D_{12}$) | Cargo throughput ($D_{13}$) | Number of takeoffs and landings in the peak hour ($D_{14}$) |
|---|---|---|---|---|
| | Most important | | | Least important |
| BO | $[1\ 1]$ | $[2\ 3]$ | $[2\ 3]$ | $[5\ 7]$ |
| OW | $[5\ 7]$ | $[3\ 5]$ | $[3\ 5]$ | $[1\ 1]$ |
| Interval weights | $[2.2134\ 2.4846]$ | $[1.0246\ 1.1954]$ | $[1.0246\ 1.1954]$ | $[0.3286\ 0.3689]$ |

**Table 2.7** The local weights of the factors in the dimension of network connectivity ($D_2$)

| | Number of navigable cities ($D_{21}$) | Percentage of transit passenger ($D_{22}$) | Flight density ($D_{23}$) | Number of airline companies ($D_{24}$) |
|---|---|---|---|---|
| | Most important | | | Least important |
| BO | $[1\ 1]$ | $[3\ 5]$ | $[2\ 4]$ | $[4\ 6]$ |
| OW | $[4\ 6]$ | $[1\ 3]$ | $[2\ 3]$ | $[1\ 1]$ |
| Interval weights | $[2.2134\ 2.8117]$ | $[0.5318\ 0.8249]$ | $[0.9899\ 1.3296]$ | $[0.4681\ 0.5946]$ |

**Table 2.8** The local weights of the factors in the dimension of service quality ($D_3$)

| | Airport security and safety ($D_{31}$) | On-time performance of flights ($D_{32}$) | Service satisfaction rate ($D_{33}$) | Logistic service ($D_{34}$) | Percentage of ground crew with college degree ($D_{35}$) |
|---|---|---|---|---|---|
| | | | Most important | | Least important |
| BO | $[1\ 1]$ | $[2\ 4]$ | $[1\ 1]$ | $[3\ 5]$ | $[6\ 8]$ |
| OW | $[6\ 8]$ | $[4\ 6]$ | $[6\ 8]$ | $[2\ 3]$ | $[1\ 1]$ |
| Interval weights | $[2.1689\ 2.2974]$ | $[1.0845\ 1.1487]$ | $[2.1689\ 2.2974]$ | $[0.5743\ 0.7230]$ | $[0.2711\ 0.2872]$ |

In a similar way, the global interval weights of all the influential factors can be determined, and the results were presented in Table 2.11.

After determining the global interval weights of these two influential factors, the elements in the probability degree matrix can be determined. Taking the probability degree of the global interval weight of airport infrastructure ($D_{11}$) be greater than that of number of navigable cities ($D_{21}$)) can be determined by Eq. (2.36), and the result was presented in Eq. (2.43).

**Table 2.9** The local weights of the factors in the dimension of operations and management ($D_4$)

| | Time for flight transfer ($D_{41}$) | Time for check-in, security check and baggage claim ($D_{42}$) | The income of main business per capita ($D_{43}$) |
|---|---|---|---|
| | Least important | Most important | |
| BO | $[3\ 5]$ | $[1\ 1]$ | $[2\ 3]$ |
| OW | $[1\ 1]$ | $[3\ 5]$ | $[1\ 3]$ |
| Interval weights | $[0.4309\ 0.6934]$ | $[1.8171\ 2.4662]$ | $[0.6934\ 1.0772]$ |

**Table 2.10** The local weights of the factors in the dimension of external environment ($D_5$)

| | Geographical conditions ($D_{51}$) | Governmental support ($D_{52}$) | GDP of the cities within the service radius of the airport ($D_{53}$) | Population of the cities within the service radius of the airport ($D_{54}$) |
|---|---|---|---|---|
| | Most important | Least important | | |
| BO | $[1\ 1]$ | $[4\ 5]$ | $[1\ 3]$ | $[3\ 5]$ |
| OW | $[4\ 5]$ | $[1\ 1]$ | $[2\ 4]$ | $[1\ 2]$ |
| Interval weights | $[1.8612\ 2.8117]$ | $[0.4162\ 0.5946]$ | $[0.9899\ 2.0000]$ | $[0.4729\ 0.8249]$ |

$$\max\left\{1-\frac{1}{2}\max\left[\frac{1.8639+3.9786-(1.7398+2.7680)}{3.9786-1.8639+2.7680-1.7398}+1,0\right],0\right\}=0.2877 \tag{2.43}$$

After determining the probability degree matrix, the relative priority of each influential factor can be determined, then these twenty factors can be ranked, as presented in Table 2.11. These twenty influential factors can be divided into three groups, including significantly important group, moderately important group, and less important group.

The significantly important group consists of the influential factors whose relative priorities are greater than 0.06, and they are geographical conditions ($D_{51}$), number of navigable cities ($D_{21}$), airport infrastructure ($D_{11}$), service satisfaction rate ($D_{33}$), airport security and safety ($D_{31}$), GDP of the cities within the service radius of the airport ($D_{53}$), and time for check-in, security check and baggage claim ($D_{42}$). The moderately important group consists of the influential factors whose relative priorities are greater between 0.045 and 0.06, and they are flight density ($D_{23}$), on-time performance of flights ($D_{32}$), passenger throughput ($D_{12}$), cargo throughput ($D_{13}$), and population of the cities within the service radius of the airport ($D_{54}$). These residual influential factors belong to the less important group.

**Table 2.11** The global interval weights of all the influential factors

| Dimensions | Factors | Global interval weights | Relative priorities | Ranking |
|---|---|---|---|---|
| Airport capacity ($D_1$) | $D_{11}$ | [1.7398 2.7680] | 0.0666 | 3 |
| | $D_{12}$ | [0.8054 1.3317] | 0.0500 | 11 |
| | $D_{13}$ | [0.8054 1.3317] | 0.0500 | 10 |
| | $D_{14}$ | [0.2583 0.4110] | 0.0280 | 19 |
| Network connectivity ($D_2$) | $D_{21}$ | [1.8639 3.9786] | 0.0703 | 2 |
| | $D_{22}$ | [0.4478 1.1673] | 0.0430 | 13 |
| | $D_{23}$ | [0.8336 1.8814] | 0.0543 | 8 |
| | $D_{24}$ | [0.3942 0.8414] | 0.0373 | 17 |
| Service quality ($D_3$) | $D_{31}$ | [1.9330 2.5138] | 0.0664 | 5 |
| | $D_{32}$ | [0.9665 1.2569] | 0.0515 | 9 |
| | $D_{33}$ | [1.9330 2.5138] | 0.0664 | 4 |
| | $D_{34}$ | [0.5118 0.7911] | 0.0377 | 16 |
| | $D_{35}$ | [0.2416 0.3142] | 0.0260 | 20 |
| Operations and management ($D_4$) | $D_{41}$ | [0.2420 0.6877] | 0.0332 | 18 |
| | $D_{42}$ | [1.0206 2.4459] | 0.0605 | 7 |
| | $D_{43}$ | [0.3895 1.0683] | 0.0409 | 14 |
| External environment ($D_5$) | $D_{51}$ | [1.6590 4.6454] | 0.0704 | 1 |
| | $D_{52}$ | [0.3710 0.9824] | 0.0392 | 15 |
| | $D_{53}$ | [0.8824 3.3043] | 0.0632 | 6 |
| | $D_{54}$ | [0.4215 1.3629] | 0.0451 | 12 |

The following implication can be obtained for selecting the location for building airports and improving the competitiveness of the airports:

1. The geographical conditions of the airport are crucial for the competitiveness of the airports, and the selection of the location of airports should give the highest priority to this issue;

2. The number of navigable cities plays a significant role for improving the competitiveness of the airports, and the larger the number, the more competitive the airport will be. In order to improve the competitiveness of the airports, the decision-makers should adopt some measures to increase the number of the navigable cities;
3. Airport infrastructure is the foundation of a competitive airport, perfecting the airport infrastructure through increasing the airport runway number, the boarding gate number, the area of the terminal building, and the number of gate positions to improve the hardware of the airports can significantly improve the competitiveness of the airports;
4. The service satisfaction rate as a key measure of service quality should be paid more attentions on this issue to improve the soft ability;
5. The airport security and safety determines the safety of the passengers and the security of the whole airport, and the violence incident and the terrorist incident which happens in the airport will significantly decrease the attraction of the corresponding airports;
6. The GDP of the cities within the service radius of the airport determines the future development and business status of the airport, and this issue should be incorporated in location selection for building new airports;
7. Time for check-in, security check and baggage claim can significantly influence the customers' choices, and the decision-makers of the airport should take some measures for reducing the time for check-in, security check and baggage claim to increase the competitiveness of airports.

## 2.5 Conclusions

In order to help the decision-makers to understand the influential factors of the competitiveness of the airports in China, this study investigated and prioritized the influential factors of China's airport. A total of twenty influential factors in five dimensions including airport capacity ($D_1$), network connectivity ($D_2$), service quality ($D_3$), operations and management ($D_4$), and external environment ($D_5$) were firstly summarized. Then, a two-stage interval best-worst method was developed for determining the relative importance of the influential factors and ranking these factors. The developed two-stage interval best-worst method has the following advantages comparing with the previous weighting method:

1. The users of this weighting method are allowed to use the interval numbers instead of crisp numbers to compare a factors with another, and the problem of vagueness, ambiguity and subjectivity existing in human's judgments can be successfully solved;
2. The proposed weighting method which is based on the best-worst method developed by Rezaei (2015; 2016) requires less times of calculation comparing with the previous weighting methods.

After the prioritization of these influential factors of airport competitiveness, some policy implications were proposed for building competitive airports and improving the competitiveness of the airports.

**Acknowledgments** This study was financially supported by The Start-up Grant of The Hong Kong Polytechnic University for New Employees (Project title: *Multi-criteria Decision Making for More Sustainable Transportation*, grant number: **1-ZE8W**).

## References

G. Alefeld, J. Herzberger, *Introduction to Interval Computation* (Academic Press, 2015)

D. An, B. Xi, J. Ren, Y. Wang, X. Jia, C. He, Z. Li, Sustainability assessment of groundwater remediation technologies based on multi-criteria decision making method. Resour. Conserv. Recy **119**, 36–46 (2017)

C.P. Barros, P. Wanke, An analysis of African airlines efficiency with two-stage TOPSIS and neural networks. J. Air Transp. Manag. **44**, 90–102 (2015)

G.C.L. Bezerra, C.F. Gomes, Measuring airport service quality: a multidimensional approach. J. Air Transp. Manag. **53**, 85–93 (2016)

C.C. Chao, P.C. Yu, Quantitative evaluation model of air cargo competitiveness and comparative analysis of major Asia-Pacific airports. Transp. Policy **30**, 318–326 (2013)

I.S. Chen, A combined MCDM model based on DEMATEL and ANP for the selection of airline service quality improvement criteria: a study based on the Taiwanese airline industry. J. Air Transp. Manag. **57**, 7–18 (2016)

Y. Chen, J.L. Peng, Research on competitiveness evaluation of the airport logistics based on AHP, in *2010 2nd International Conference on Information Science and Engineering (ICISE)* (IEEE, 2010), pp. 483–486

T.W. Chung, J.K. Han, Evaluating competitiveness of transshipment Cargo in major airports in Northeast Asia: airport branding. Asian J. Shipp. Logist. **29**(3), 377–394 (2013)

A.R. Correia, S.C. Wirasinghe, A.G. de Barros, A global index for level of service evaluation at airport passenger terminals. Transp. Res. E-Log **44**(4), 607–620 (2008)

Q. Cui, C. Wu, H. Kuang, Influencing factors research of airports competitiveness based BP-DEMATEL model. Syst. Eng.-Theor. Pract. **33**(6), 1471–1478 (2013a). (in Chinese)

Q. Cui, H.B. Kuang, C.Y. Wu, Y. Li, Dynamic formation mechanism of airport competitiveness: the case of China. Transp. Res. Part A Policy Pract **47**, 10–18 (2013b)

Q. Cui, Y.M. Wei, Y. Li, W.X. Li, Exploring the differences in the airport competitiveness formation mechanism: evidence from 45 Chinese airports during 2010–2014. Transportmetrica B **5**(3), 330–346 (2017)

Z. Dong, Y. Peng, T. Cui, An analysis of the characters of Chinese Midwest airports and the evaluation of airport's competence. J. Beijing Inst. Technol. (Soc. Sci. Edn.) **19**(3), 82–87 (2007). (in Chinese)

C.P. Garg, A robust hybrid decision model for evaluation and selection of the strategic alliance partner in the airline industry. J. Air Transp. Manag. **52**, 55–66 (2016)

C.C. Hsu, J.J. Liou, An outsourcing provider decision model for the airline industry. J. Air Transp. Manag. **28**, 40–46 (2013)

X. Liang, J. Li, J. Liu, H. Qiao, H. Yin, Study on the multilayer airport system in China based on evaluation of airport competitiveness, **28**(12), 116–126 (2016) (in Chinese)

T. Lupo, Fuzzy ServPerf model combined with ELECTRE III to comparatively evaluate service quality of international airports in Sicily. J. Air Transp. Manag. **42**, 249–259 (2015)

M.M. Pandey, Evaluating the service quality of airports in Thailand using fuzzy multi-criteria decision making method. J. Air Transp. Manag. **57**, 241–249 (2016)
Y. Park, Application of a fuzzy linguistic approach to analyse Asian airports' competitiveness. Transp. Plan Technol. **20**(4), 291–309 (1997)
Y. Park, An analysis for the competitive strength of Asian major airports. J. Air Transp. Manag. **9**(6), 353–360 (2003)
Y. Peng, Y. Li, Evaluation of Hub Airports' competence. Technoecon. Manag. Res. **9**, 11–15 (2011). (in Chinese)
J. Rezaei, Best-worst multi-criteria decision-making method. Omega **53**, 49–57 (2015)
J. Rezaei, Best-worst multi-criteria decision-making method: some properties and a linear model. Omega **64**, 126–130 (2016)
T.L. Saaty, Exploring the interface between hierarchies, multiple objectives and fuzzy sets. Fuzzy Sets Syst. **1**(1), 57–68 (1978)
D. Shen, C. Yu, Research on the logistics competitiveness evaluation of airport based on FCE-AHP. Mod. Manag. (Chinese) **2**, 51–57 (2012)
D. Su, Construction of the airport competitiveness evaluation index system. J. Civil Aviat. Flight Univ. China **22**(1), 18–22 (2011). (in Chinese)
D. Su, M. Wang, G. Zhou, The construction of evaluation index system and the competitiveness of airports. J. Hunan Finan. Econ. Coll. **26**(123), 98–100 (2010). (in Chinese)
Y.M. Wang, J.B. Yang, D.L. Xu, A two-stage logarithmic goal programming method for generating weights from interval comparison matrices. Fuzzy Sets Syst. **152**(3), 475–498 (2005)
F. Xie, H. Xia, M. Wang, M. Wu, The evaluation model of airport competitiveness based on the perceptions and satisfactions of the consumers. Transp. Inf. Saf. **33**(3), 40–46 (2015). (in Chinese)
Z. Xu, Dependent uncertain ordered weighted aggregation operators. Inf. Fus. **9**(2), 310–316 (2008)
Z. Xu, Q. Da, A possibility-based method for priorities of interval judgment matrices. Chin. J. Manage. Sci. **11**(1), 63–65 (2003). (in Chinese)
C.H. Yeh, Y.L. Kuo, Y.H. Chang, Fuzzy multiattribute evaluation of airport performance, in *2011 IEEE International Conference on Fuzzy Systems (FUZZ)* (IEEE, 2011), pp. 2630–2637
G.T. Yeo, Y. Wang, C.C. Chou, Evaluating the competitiveness of the aerotropolises in East Asia. J. Air Trans. Manag. **32**, 24–31 (2013)
D. Zietsman, M. Vanderschuren, Analytic Hierarchy process assessment for potential multi-airport systems–the case of Cape Town. J. Air Transp. Manag. **36**, 41–49 (2014)
W. Zhao, The competitiveness of global airports and analysis of the development of China's hub airports. Airport Transp. Bus. **374**(7), 24–31 (2016). (in Chinese)
L.G. Zhou, H.Y. Chen, J.M. Merigó, A.M. Gil-Lafuente, Uncertain generalized aggregation operators. Expert Syst. Appl. **39**(1), 1105–1117 (2012)

# Chapter 3
# 2-Tuple DEMATEL for Complex Interrelationships Analysis: Barriers Identification, Cause-Effect Analysis and Policy Implications for Sustainable Tourism Industry

**Abstract** Tourism industry plays an important role in China's economy growth; however, there are many interviewed and interacted barriers hindering the development of China's tourism industry. The objective of this study is to investigate the barriers hindering the development of China's tourism industry, develop an improved DEMATEL method to study the cause-effect relationships among these barriers and identify the critical barriers according to their relative importance. Ten barriers including weak industrial foundation, leisure consciousness, lack of policy support, lack of complete infrastructure, staff quality, lack collaborations among tourism enterprises, lack of innovations, governance and management ability, weak tourism brand, incomplete legal and regulation system, and insufficient governmental investment, were studied in this paper, and the 2-tuple DEMATEL was employed to analyze the complex relationships among these barriers and rank these barriers, and some policy implications were presented to China's decision-makers for them to promote the development of China's tourism industry.

## 3.1 Introduction

As the fourth largest export industry, tourism industry plays an important role in economy growth all over the world, because it is a multi-aspect industry which consists of food and hospitality, transportation, accommodation, victor attractions, and retail, etc. (Morrison and Pickering 2013). However, there are more and more requirements for tourism industry nowadays. For instance, Global tourism industry leaders realize that sustainable tourism development should be implemented for the conservation of nature and the preservation of indigenous culture (Hassan 2000). Low-carbon tourism for mitigating climate change has also attracted more and more attentions (Dwyer et al. 2013; Tang et al. 2014). Besides these new requirements, the tourism industry in some countries also faces many barriers for its further development. Taking China as an example, it is the second largest economy in the world, China's tourism industry has made significant progress after China's opening-up policy, the total visitor arrivals to China accounted to 135 million compared with 1.8 million in 1978 (Leung et al. 2014). However, Bao et al. (2014) pointed out that most research about tourism focused on applied work whereas little priority has



J. Ren, *Advanced Operations Management for Complex Systems Analysis*, SpringerBriefs in Applied Sciences and Technology,
https://doi.org/10.1007/978-3-030-45418-0_3

been given to theory development. Meanwhile, there are also some other barriers that hinder the development of China's tourism industry, i.e. low level of standardization and normalization, slow progress in brand diversity, lack skilled human resource, and incomplete regulatory system, etc. (Chen and Ni 2001; Li 2013). Therefore, there is a debate on how to promote the development of China's tourism industry.

These barriers hindering the development of China's tourism industry are not independent but interacted and intertwined, thus, it is usually difficult for the decision-makers in China to have a good understanding of the cause-effect relationships among these barriers and provide effective measures to solve these problems. Moreover, there are multiple barriers with interdependences and interactions, thus, the analysis of the barriers existed in China's tourism industry usually involves multiple criteria, and multi-criteria decision-making (MCDM) methods are usually used to address this kind of problems.

The purpose of this study is to investigate the barriers hindering the development of China's tourism industry, develop an easy-operation MCDM method to study the cause-effect relationships among these barriers, identify the critical barriers, and provide effective strategic measures to overcome these barriers for promoting the development of China's tourism industry. Besides this part, the remaindering parts of this study have been organized as follows: Sect. 2.2 has a comprehensive overview of the MCDM methods for tourism management; Sect. 2.3 introduces the MCDM method; Sect. 2.4 presents the results and discussion; and finally, policy implications and conclusion are presented in Sect. 2.5.

## 3.2 Literature Review

MCDM methods which has the ability for ranking the alternatives with the consideration of multiple criteria have been widely used in tourism management. Zhang et al. (2011) employed Technique for Order Preference by Similarity to Ideal Solution (TOPSIS) and information entropy to evaluate the tourism destination competitiveness. Alptekin and Büyüközkan (2011) developed an integrated case-based reason and MCDM technique for Web based tourism destination planning. Liu et al. (2013) combined DEMATEL, DEMATEL-based ANP and VIKOR to investigate the influential relationships among dimensions and criteria for proposing the improvement schemes for the connection service between metro system and urban airport. Tsaur and Wang (2007) developed a multi-criteria decision making method by combining Analytic Hierarchy Process (AHP) and fuzzy set theory to evaluate the sustainable tourism development. Jeon and Kim (2011) combined SWOT analysis and AHP to evaluate the critical factors in strategic planning and develop effective strategies for Chuncheon, a tourist destination in South Korea.

MCDM methods for analyzing the cause-effect relationships among multiple criteria such as Decision making trial and evaluation laboratory (DEMATAL) has also been widely used in tourism management. The Science and Human Affairs Program of the Battelle Memorial Institute of Geneva developed the DEMATEL method between 1972 and 1976 for providing an effective method to study the complicated and intertwined problem group (Gabus and Fontela 1973; Fontela and Gabus 1976). The foundation of this method is graph theory, and it has the ability to investigate the cause-effect relationships among the factors in the complicated and intertwined problem group and identify the critical factors (Hu et al. 2011; Ho et al. 2011). DEMATEL has been widely used in tourism management. For instance, Tsai et al. (2010) employed the DEMATE method to detect complex relationships and build a network structure among the cost and differentiation advantage criteria of the international tourism hotel. Chen (2012) used the DEMATEL method to study the eleven influencing factors in five clusters (i.e., the strengthening of infrastructure, the strengthening of tourist services, the clarity of market segmentation, marketing planning and government policy) for finding the strategy for developing the medical tourism in Taiwan. Liu et al. (2012) used DEMATEL to construct a network relationship map to illustrate the influential network of the tourism policy for improving the tourism policy implementation. Chen et al. (2011) combined DEMATEL and Analytic Network Process (ANP) to help the business managers to understand the appropriate actions for the management of the hot spring hotels. Tseng (2009) developed a modified DEMATEL method by combing the fuzzy set theory and the traditional DEMATEL to analyze the cause-effect relationship among the influencing factors of hotel service quality.

The traditional DEMATEL is to study the cause-effect relationships among various influencing factors of the complex system by using crisp numbers like 0, 1, 2, 3, 4 and 5 to represent the relative influences of one factor on another; however, it is usually difficult for the decision-makers to use crisp numbers to express their opinions due to the imprecision and vagueness existed in human's judgments (Tseng 2009). Fuzzy set theory has the ability to address the vagueness, ambiguity and subjectivity existed in human's judgments, thus, it has been combined with the traditional DEMATEL method in many studies (Ren and Sovacool 2014; Liu et al. 2015; Kabak et al. 2015). Among these, the DEMATEL methods combined with 2-tuple fuzzy linguistic representation model for computing with words which is composed by a linguistic term and a numeric value, received more and more attentions for the advantages of overcoming the loss of information caused by the need to express the results in the initial expression domain (Herrera and Martínez 2000). Therefore, the 2-tuple DEMATEL technique was applied in this study to investigate the cause-effect relationships among the influencing factors of China's tourism industry.

## 3.3 Methods

This section consists of three parts, Sect. 3.3.1 is about the identification of the barriers that hinder the development of China's tourism industry; Sect. 3.3.2 introduced the definitions and operations of 2-tuple linguistic variables; and Sect. 3.3.3 introduce the improved DEMATEL method.

### 3.3.1 Barriers Identification

In order to obtain the barriers that hinder the development of China's tourism industry, all the papers related China's tourism industry by searching the keywords "China", "tourism", "barriers" in Google Scholar and China National Knowledge Infrastructure were obtained, the most representative papers such as Zhou (2009), Yuan and Li (2011), Chen and Ni (2001) were presented to the experts when holding the focus group meeting. A total of eleven barriers were identified, they are:

(1) Weak industrial foundation ($B_1$): China's tourism industry started late compared with the developed western countries, and the industrial foundation is weak.
(2) Leisure consciousness ($B_2$): The leisure consciousness of China's citizens is weak as the basic necessities of life for them in the traditional mind are food, clothing, shelter and transportation excluding tourism, and it is difficult to change this old mind even the economy becomes better and better.
(3) Lack of policy support ($B_3$): Compared with some other industries, China's tourism lacks enough governmental support, i.e. subsidies, and low interest or zero interest loans.
(4) Lack of complete infrastructure ($B_4$): Some beauty spots are located in the outskirts, and the infrastructure, especially the transportation infrastructure, is incomplete for the further development of China's tourism industry.
(5) Staff quality ($B_5$): Staff quality mainly refers to the education background and professional skills of the staff in the tourism industry, both of these influence the service quality.
(6) Lack collaborations among tourism enterprises ($B_6$): The tourism enterprises in China are usually isolated without any collaboration.
(7) Lack of innovations ($B_7$): The tours operated by the tourism companies are always old traditional ones; however, these cannot satisfy the requirements of new-generation tourists.
(8) Governance and management ability ($B_8$): The level of management also has significant impacts on the development of China's tourism industry.
(9) Weak tourism brand ($B_9$): There are various consume disputes appeared in China's media, and these have serious negative impacts on the tourism brand of some China's beauty spots.
(10) Incomplete legal and regulation system ($B_{10}$): The corresponding legislations, regulations and laws with respect to China's tourism industry are incomplete.
(11) Insufficient governmental investment ($B_{11}$): China's administration has not invested sufficiently on its tourism industry as one pillar of the third industry.

### 3.3.2 *Definitions and Operations of 2-Tuple Linguistic Variables*

The definitions of 2-tuple linguistic variables and the operations between/among 2-tuple linguistic variable were presented as follows (Chen and Tai 2005; Herrera and Martínez 2000):

**Definition 3.1** Let $S = \{s_0, s_1, \ldots, s_g\}$ be a linguistic term set and $\beta \in [0, 1]$ a value representing the result of a symbolic aggregation operation, and the generated translation function $\Delta$ can be used to obtain the 2-tuple linguistic variable equivalent to $\beta$, as defined in Eqs. (3.1) and (3.2).

$$\Delta : [0, 1] \rightarrow \left[-\frac{1}{2g}, \frac{1}{2g}\right) \tag{3.1}$$

$$\Delta(\beta) = (s_i, \alpha), \, with \begin{cases} s_i, \; i = round(\beta.g) \\ \alpha = \beta - \dfrac{i}{g}, \;\; \alpha \in \left[-\frac{1}{2g}, \frac{1}{2g}\right) \end{cases} \tag{3.2}$$

where $round(\cdot)$ is the usual rounding operation, $s_i$ has the closest index label to $\beta$, and $\alpha$ is the value of the symbolic translation. The interval $\left[-\frac{1}{2g}, \frac{1}{2g}\right)$ is determined by the number of linguistic terms in S. For instance, if S consists of five linguistic terms, then $g = 5$ and $\alpha \in [-0.1, 0.1)$.

**Definition 3.2** Let $S = \{s_0, s_1, \ldots, s_g\}$ be a linguistic term set and $(s_i, \alpha)$ be a 2-tuple. The reverse function $\Delta^{-1}$ can transform a 2-tuple linguistic variable into its equivalent numerical value $\beta \in [0, 1]$, as defined in Eqs. (3.3) and (3.4).

$$\Delta^{-1} : S \times \left[-\frac{1}{2g}, \frac{1}{2g}\right) \rightarrow [0, 1] \tag{3.3}$$

$$\Delta^{-1}(s_i, \alpha) = \frac{i}{g} + \alpha = \beta \tag{3.4}$$

**Definition 3.3** Let $(s_k, \alpha_1)$ and $(s_l, \alpha_2)$ be two 2-tuple, then,

(1) If $k < l$, then $(s_k, \alpha_1)$ is smaller than $(s_l, \alpha_2)$;
(2) If $k = l$, then

a. If $\alpha_1 = \alpha_2$, then $(s_k, \alpha_1)$ is equal to $(s_l, \alpha_2)$;
b. If $\alpha_1 < \alpha_2$, then $(s_k, \alpha_1)$ is smaller than $(s_l, \alpha_2)$; and
c. If $\alpha_1 > \alpha_2$, then $(s_k, \alpha_1)$ is greater than $(s_l, \alpha_2)$.

**Definition 3.4** Let $X = \{(s_1, \alpha_1), (s_2, \alpha_2), \ldots, (s_n, \alpha_n)\}$ be a set of 2-tuples, and their arithmetic mean $\bar{X}$ is computed as

$$\bar{X} = \Delta\left(\frac{1}{n}\sum_{j=1}^{n}\Delta^{-1}(s_j, \alpha_j)\right) = \Delta\left(\frac{1}{n}\sum_{j=1}^{n}\beta_j\right) \tag{3.5}$$

**Definition 3.5** Let $X = \{(s_1, \alpha_1), (s_2, \alpha_2), \ldots, (s_n, \alpha_n)\}$ be a set of 2-tuples and $W = (\omega_1, \omega_2, \ldots, \omega_n)^T$ be their associate weights, with $\omega_j \in [0, 1]$, $j = 1, 2, \ldots, n$, $\sum_{i=1}^{n}\omega_j = 1$. The 2-tuple weighted average (TWA) is defined as

$$TWA(X) = \Delta\left(\sum_{j=1}^{n}\omega_j\Delta^{-1}(s_j, \alpha_j)\right) = \Delta\left(\sum_{j=1}^{n}\omega_j\beta_j\right) \tag{3.6}$$

### *3.3.3 Improved DEMATEL*

The developed DEMATEL consists of five main steps based on the previous works (Gabus and Fontela 1973; Fontela and Gabus 1976; Ren et al. 2013; Liang et al. 2016):

(1) Determining the direct-influenced matrices using linguistic terms.
(2) Determining the weighted direct-influenced matrix.
(3) Normalizing the initial direct-relation matrix.
(4) Calculating the total relation matrix.
(5) Drawing the cause-effect relationships diagram.

These five steps are specified as follow:

**Step 1**: Determining the direct-influenced matrices using linguistic terms. The objective of this step is to determine the relative degree of one factor affecting another factor by using the linguistic terms in the linguistic term set S = {$s_0$ = "No influence (N)", $s_1$ = "Very low influence (VL)", $s_2$ = "Low influence (L)", $s_3$ = "Medium influence (M)", $s_4$ = "High influence (H)", and $s_5$ = "Very high influence (VH)"}. According to Definitions 3.1 and 3.2, g = 5, $\alpha \in [-0.1, 0.1)$, and $\Delta^{-1}(s_0, 0) = 0$, $\Delta^{-1}(s_1, 0) = 0.2$, $\Delta^{-1}(s_2, 0) = 0.4$,$\Delta^{-1}(s_3, 0) = 0.6$,$\Delta^{-1}(s_4, 0) = 0.8$, and $\Delta^{-1}(s_5, 0) = 1.0$. The participants are asked to evaluate the effect of a factor on another factor using a 2-tuple which corresponds to a linguistic term. For instance, if the participants think that the relative effect of one factor on another factor is "medium influence", denoted by $(s_3, 0)$.

Assuming there are H participants and n factors to be considered. The influence of the i-th factor on the j-th factor determined by the $k$-th participant denotes by $\tilde{x}_{ij}^{k}$ which is a 2-tuples. The results provided by the k-th participant can form a n × n matrix, as shown in Eqs. (3.7) and (3.8):

$$X^k = \begin{bmatrix} (s_0, 0) & \tilde{x}_{12}^k & \cdots & \tilde{x}_{1n}^k \\ \tilde{x}_{21}^k & (s_0, 0) & \cdots & \tilde{x}_{2n}^k \\ \vdots & \vdots & \ddots & \vdots \\ \tilde{x}_{n1}^k & \tilde{x}_{n2}^k & & (s_0, 0) \end{bmatrix}, k = 1, 2, \ldots, H \tag{3.7}$$

$$\tilde{x}_{ij}^k = \left(s_{ij}^k, 0\right), \tilde{x}_{ij}^k \in S \tag{3.8}$$

where $X^k$ represents the direct-influenced matrix determined by the $k$-th expert.
**Step 2**: Determining the weighted direct-influenced matrix. The weighted direct-influenced matrix, so-called "initial direct-relation matrix", could be calculated by Eq. (3.9).

$$A = \left[\tilde{a}_{ij}\right]_{n\times n} = \begin{bmatrix} (s_0, 0) & \tilde{a}_{12} & \cdots & \tilde{a}_{1n} \\ \tilde{a}_{21} & (s_0, 0) & \cdots & \tilde{a}_{2n} \\ \vdots & \vdots & \ddots & \vdots \\ \tilde{a}_{n1} & \tilde{a}_{n2} & & (s_0, 0) \end{bmatrix} \tag{3.9}$$

However, it is worth pointing out that there are H participants who represent different stakeholders, and the relative importance of their roles is different. Therefore, the 2-tuple weighted average (TWA) method is employed to determine the average effect of one criterion on another. According to the Definition 3.5, the elements in the matrix A can be determined. Note that, if the relative importance of their roles is equal, and Eq. (3.10) can be transformed into Eq. (3.11) which is the arithmetic mean of the 2-tuples provided by the H experts in Step 1.

$$\tilde{a}_{ij} = \Delta\left(\sum_{k=1}^{H} \omega_k \Delta^{-1}\left(x_{ij}^k, 0\right)\right) \tag{3.10}$$

$$\tilde{a}_{ij} = \Delta\left(\sum_{k=1}^{H} \frac{1}{H} \Delta^{-1}\left(x_{ij}^k, 0\right)\right) \tag{3.11}$$

where A is the average matrix.

Then, the 2-tuple $\tilde{a}_{ij}$ in the weighted direct-influenced matrix can be transformed into its equivalent numerical value $a_{ij}$ by Eq. (3.4).

In this step, the relative importance of these participants can be determined by the best-worst method, and it consists of four sub-steps (Rezaei 2015a, b):

**Sub-step 1**: Determining the most important and the least stakeholders, denotes by $C_M$ and $C_L$, respectively.
**Sub-step 2**: Determining the relative preferences of the most important stakeholder over all the other stakeholders and that of all the other stakeholders over the least important stakeholder by using the scales used on Saaty method (Saaty

**Table 3.1** Comparison scale in Saaty method (Saaty 1980)

| Scales | Definition | Note |
|---|---|---|
| 1 | Equal importance | $i$ is equally important to $j$ |
| 3 | Moderate importance | $i$ is moderately important to $j$ |
| 5 | Essential importance | $i$ is essentially important to $j$ |
| 7 | Very strong importance | $i$ is very strongly important to $j$ |
| 9 | Absolute importance | $i$ is very absolutely important to $j$ |
| 2, 4, 6, 8 | Intermediate value | The relative importance of $i$ to $j$ is between to adjacent judgment |
| Reciprocal | Reciprocals of above | The value had been assigned to $i$ when compared to $j$, then $j$ has the reciprocal value compared to $i$ |

1980), see Table 3.1. Then, the Most-to-Others (MO) vector and the Others-to- Least (OL) vector can be obtained, as presented in Eqs. (3.12) and (3.13), respectively.

$$MO = \left[ a_{M1} \; a_{M2} \; \ldots \; a_{Mn} \right] \tag{3.12}$$

$$OL = \left[ a_{1W} \; a_{2W} \; \ldots \; a_{nW} \right] \tag{3.13}$$

where $a_{Mj}(j = 1, 2, \ldots, n)$ and $a_{jL}(j = 1, 2, \ldots, n)$ represent the relative preference of the most important stakeholder over the $j$-th stakeholder and the $j$-th stakeholder over the least important stakeholder.

It is apparent that when $j = M$, then $a_{Mj} = 1$, and when $j = L$, then $a_{jL} = 1$.

**Sub-step 3**: Determining the weights of the criteria.

The optimal weights of the criteria should satisfy the conditions presented in Eqs. (3.14) and (3.15).

$$\frac{\omega_M}{\omega_j} = a_{Mj}(j = 1, 2, \ldots, n) \tag{3.14}$$

$$\frac{\omega_j}{\omega_L} = a_{jL}(j = 1, 2, \ldots, n) \tag{3.15}$$

$$\begin{aligned} & \min \max_j \left\{ \left| \frac{\omega_M}{\omega_j} - a_{Bj} \right|, \left| \frac{\omega_j}{\omega_L} - a_{jL} \right| \right\} \\ & s.t. \\ & \sum_{j=1}^{n} \omega_j = 1 \\ & \omega_j \geq 0, j = 1, 2, \ldots, n \end{aligned} \tag{3.16}$$

Programming (3.16) can be transferred into the following problem:

$$
\begin{aligned}
&\min \xi \\
&s.t. \\
&\left| \frac{\omega_M}{\omega_j} - a_{Mj} \right| \le \xi, j = 1, 2, \ldots, n \\
&\left| \frac{\omega_j}{\omega_L} - a_{jL} \right| \le \xi, j = 1, 2, \ldots, n \\
&\sum_{j=1}^{n} \omega_j = 1 \\
&\omega_j \ge 0, j = 1, 2, \ldots, n
\end{aligned}
\tag{3.17}
$$

where $\omega_M$ represents the weight of the most important stakeholder, $\omega_L$ represent the weight of the least important stakeholder, and $\omega_j$ denotes the weight of the $k$-th stakeholder.

The $\xi^*$ is the value of the objective function in programming (3.17) under the optimum conditions $\omega_1^*$,$\omega_2^*$,..., and $\omega_H^*$.

**Sub-step 4**: Consistency check.

The comparison is fully consistent when $a_{Bj} a_{jW} = a_{BW} (j = 1, 2, \ldots, n)$, however this ideal condition cannot always be achieved due to the ambiguity and vagueness existed in human judgments. The consistency ratio can be calculated for consistency check, as presented in Eq. (3.18),

$$
CR = \frac{\xi^*}{CI} \tag{3.18}
$$

where CR represents the consistency ratio, and CI represents the consistency index.

The consistency index can be obtained according to Table 3.2, and the value of consistency ratio belonging to the interval $\left[0\ 1\right]$ indicates the consistency level, and the closer the value to zero, the more consistent the comparison is; on the contrary, the closer the value to one, the more consistent the comparison is.

**Table 3.2** Consistency Index (CI) table (Rezaei 2015a, b)

| $a_{ML}$ | 1 | 2 | 3 | 4 | 5 | 6 | 7 | 8 | 9 |
|---|---|---|---|---|---|---|---|---|---|
| Consistency index (max $\xi$) | 0.00 | 0.44 | 1.00 | 1.63 | 2.30 | 3.00 | 3.73 | 4.47 | 5.23 |

**Step 3**: Normalizing the initial direct-relation matrix. The normalized initial direct-relation matrix D could be obtained by Eqs. (3.19)–(3.21).

$$s = \max\left\{\left(\max_{1\le i\le n}\sum_{j=1}^{n}\Delta^{-1}\left(\tilde{a}_{ij}\right)\right), \left(\max_{1\le j\le n}\sum_{i=1}^{n}\Delta^{-1}\left(\tilde{a}_{ij}\right)\right)\right\} \tag{3.19}$$

$$D = \left[d_{ij}\right]_{n\times n} \tag{3.20}$$

$$d_{ij} = \frac{1}{s}\times\Delta^{-1}\left(\tilde{a}_{ij}\right) = \frac{\Delta^{-1}\left(\tilde{a}_{ij}\right)}{s} \tag{3.21}$$

where $D$ is the normalized initial direct-relation matrix, s represents the biggest value among the sums of each row and each column, and $d_{ij}$ is the element in cell $(i, j)$ in the matrix $D$.

The sum of each row in matrix $D$ represents the direct effects of the corresponding factor on the other factors, and $\max\limits_{1\le i\le n}\sum_{j=1}^{n}\Delta^{-1}\left(\tilde{a}_{ij}\right)$ represents the factor has the highest total influence on other factors. On the other hand, the sum of each column represents the direct influences on the factor affected by the other factors, and $\max\limits_{1\le j\le n}\sum_{i=1}^{n}\Delta^{-1}\left(\tilde{a}_{ij}\right)$ represents the factor which is the most influenced by the other factors.

**Step 4**: Calculating the total relation matrix (Gabus and Fontela 1973; Fontela and Gabus 1976). The total relation matrix can be calculated by summing the direct effects which is expressed in D and all the indirect effects which can be determined by raising D to different powers. A continuous decrease of the indirect effects of factors along the powers of matrix $D$, such as $D^2$, $D^3$, …, $D^\infty$, similar to Markov chain matrix, guarantees convergent solutions to the matrix inversion. Then, the total relation matrix T could be calculated by Eqs. (3.22) and (3.23).

$$T = \left[t_{ij}\right]_{n\times n} = D^1 + D^2 + \cdots + D^h, h\to\infty, D^h = \{0\}_{n\times n} \tag{3.22}$$

$$T = D(I - D)^{-1} \tag{3.23}$$

where $T = \left\{t_{ij}\right\}_{n\times n}, D = \left\{d_{ij}\right\}_{n\times n}$, and $T$ represents the total relation matrix and I is the identity matrix.

The total effects that directly and indirectly exerted by the $i$-th factor, is denoted by $r_i$, could be calculated by Eq. (3.24).

$$r_i = \sum_{j=1}^{n} t_{ij} \tag{3.24}$$

The total effect including direct and indirect effects received by the j-th factor, and denoted by $c_j$ could be calculated by Eq. (3.25).

$$c_j = \sum_{i=1}^{n} t_{ij} \tag{3.25}$$

Therefore, when $i = j$, the sum $(r_i + c_i)$ represents the total effects given and received by the $i$-th factor, and it is a measure of the relative importance of the $i$-th factor in the complex system; the difference $(r_i - c_i)$ called "relation" shown the net effect that contributed by the $i$-th factor to the system. In the traditional DEMATEL method, the $i$-th factor is a net cause, it belongs to the 'cause group' when $(r_i - c_i) > 0$, while $(r_i - c_i) < 0$, the i-the factor is a net receiver or result, it belongs to the 'effect group'.

**Step 5**: Drawing the cause-effect relationships diagram and determining the relative importance of these factors.

After determining the coordinate values $(r_i + c_i, r_i - c_i)$ with respect to all the factors, and they could be drawn out in the cause-effect relationship diagram.

As to the relative importance of these factors, there are usually two ways: one is to determine the relative weights of the factors according to the value of $(r_i + c_i)$ with respect to the $i$-th factor, another is to determine the relative weights of the factors by Eq. (3.26) (Liu et al. 2015). The second method by using the causal diagram to set the weights of the factors (Dalalah et al. 2011; Baykasoğlu et al. 2013) was applied in this study, and the normalized weights which represent the relative importance of these factors can be determined by Eq. (3.27).

$$\bar{\omega}_j = \sqrt{(r_j + c_j)^2 + (r_j - c_j)^2} \tag{3.26}$$

$$\omega_j = \frac{\bar{\omega}_j}{\sum_{j=1}^{n} \omega_j} \tag{3.27}$$

If the factor $i$ belongs to the 'effect' group, and the value of $r_i + c_i$ represents the degree of the relative importance of the factor $i$ as one of the core factors that require to be solved, while it is not the origin of the problem. If the factor i belong to the 'cause' group, and the value of $r_i + c_i$ represents the degree of the relative importance of the factor i as one of the core driving factor for solving the key problems existed in the complex problem.

## 3.4 Results and Discussions

In order to comprehensively investigate the opinions of all the stakeholders, four groups of stakeholders were invited to participate in determining the direct-influenced matrices, including the tourist group, tour guide group, manager group, and researcher group. The tourist group consists of six tourists who usually travel in China, the tour guide group consists of four tour guides who worked in the travel companies, the manager group consists of four managers who worked in the management position of the travel companies for many years, and the research group includes seven researchers (2 professor in management science, 2 PhD student in tourism management, and 1 senior research in operations research). Four focus group meetings organized by the authors were held to collect the opinions of the experts in these four groups, and each coordinator was nominated in each group meeting. It is worth pointing out that there are usually debates among the experts in each group, and the coordinator of each group is responsible to coordinate for achieving a final consensus. The results of the direct-influenced matrices determined by the experts of these four groups were presented in Table 3.1. For instance, there are four linguistic terms including N, VL, VL, VL in cell $(1, 2)$ of Table 3.3, and it means that the effect of the weak industrial foundation ($B_1$) on the leisure consciousness ($B_2$) determined by the tourist group (group 1), tour guide group (group 2), manager group (group 3), and researcher group (group 4) are 'no influence', 'very low influence', 'very low influence', and 'very low influence', respectively.

In order to obtain the weighted direct-influenced matrix, the relative weights of the stakeholders which represent the relative importance of the stakeholders can be determined by the BW method, and the four steps of the BW method were presented as follows:

**Sub-step 1**: The most important and the least important stakeholders are manager group and tourist group, respectively among these four groups including the tourist group (group 1), tour guide group (group 2), manager group (group 3), and researcher group (group 4).

**Sub-step 2**: The relative preferences of the most important stakeholder group (manager group) over all the other and that of all the other groups over the least important stakeholder (tourist group) were also determined, as presented in Eqs. (3.28) and (3.29).

t

$$BO = \left[ a_{31} \; a_{32} \; a_{33} \; a_{34} \right] = \left[ 7 \; 3 \; 1 \; 2 \right] \tag{3.28}$$

$$OW = \left[ a_{11} \; a_{21} \; a_{31} \; a_{41} \right] = \left[ 1 \; 2 \; 7 \; 4 \right] \tag{3.29}$$

**Step 3**: Determining the optimal weights of the four stakeholder groups by solving the following programming (3.30)

**Table 3.3** The direct-influenced matrices by using linguistic terms

| | $B_1$ | $B_2$ | $B_3$ | $B_4$ | $B_5$ | $B_6$ | $B_7$ | $B_8$ | $B_9$ | $B_{10}$ | $B_{11}$ |
|---|---|---|---|---|---|---|---|---|---|---|---|
| $B_1$ | N | N, VL, VL, VL | L, N, VL, N | M, M, H, H | N, N, VL, N | H, M, L, M | H, H, VH, H | N, N, N, N | VH, H, VH, VH | VL, L, N, VL | H, VH, M, H |
| $B_2$ | N, N, VL, N | N | L, L, VL, L | M, VL, L, VL | N, N, L, VL | N, N, N, N | L, M, L, M | N, N, N, VL | VL, VL, N, N | L, L, L, VL | M, VL, L, L |
| $B_3$ | VH, H, VH, M | M, M, VH, M | N | H, H, VH, H | M, L, L, M | M, H, H, L | M, L, VH, H | VL, M, M, L | M, M, M, H | N, VL, L, VL | VH, VH, L, M |
| $B_4$ | VH, VH, VH, VH | VH, M, H, H | VL, VL, N, N | N | N, N, N, N | M, L, L, L | H, L, M, H | L, L, M, M | VH, M, VH, H | N, N, VL, N | N, N, N, N |
| $B_5$ | M, VL, VL, L | N, N, VL, VL | N, N, N, N | N, N, N, N | N | M, M, M, M | VH, VH, VH, H | M, VH, VH, H | VH, H, H, M | N, N, N, N | N, VL, N, N |
| $B_6$ | H, M, VH, H | N, N, N, VL | VL, L, VL, N | L, L, L, VL | VL, VL, N, L | N | L, VL, M, VL | M, H, L, VL | VH, VH, M, H | N, VL, N, N | VL, L, VL, VL |
| $B_7$ | H, M, H, H | N, N, N, N | N, N, N, N | N, N, VL, VL | N, N, N, N | VH, VH, VH, VH | N | M, VL, VL, VL | H, H, VH, VH | N, N, N, VL | VL, VL, N, N |
| $B_8$ | L, L, VL, N | N, N, N, N | N, N, N, N | N, N, N, N | VL, L, M, L | M, L, M, M | VL, M, H, M | N | M, M, H, H | N, VL, N, VL | L, N, VL, N |
| $B_9$ | L, M, VL, L | M, H, H, L | L, L, L, VL | N, VL, N, N | N, N, N, N | N, L, N, VL | N, VL, L, VL | N, N, N, VL | N | N, N, N, VL | VL, N, VL, N |
| $B_{10}$ | VH, H, VH, VH | VL, VL, L, M | M, VL, VL, VL | N, N, VL, N | M, M, VL, L | M, H, M, L | L, M, L, H | M, M, M, L | H, H, M, H | N | Vl, N, N, N |
| $B_{11}$ | VH, VH, VH, VH | VL, VL, L, VL | N, VL, N, N | VH, VH, VH, H | VL, VL, M, L | M, H, M, H | M, L, M, L | N, N, N, VL | H, M, M, H | N, N, N, N | N |

$$
\begin{aligned}
&\min \xi \\
&s.t. \\
&\left|\frac{\omega_3}{\omega_1}-7\right| \leq \xi \\
&\left|\frac{\omega_3}{\omega_2}-3\right| \leq \xi \\
&\left|\frac{\omega_3}{\omega_4}-2\right| \leq \xi \\
&\left|\frac{\omega_2}{\omega_1}-2\right| \leq \xi \\
&\left|\frac{\omega_4}{\omega_1}-4\right| \leq \xi \\
&\sum_{j=1}^{4} \omega_j = 1 \\
&\omega_j \geq 0, j = 1,2,\ldots,4
\end{aligned}
\tag{3.30}
$$

The results are: $\xi = 0.0202$, $\omega_1 = 0.0707$, $\omega_2 = 0.1616$, $\omega_3 = 0.5051$, $\omega_4 = 0.2626$.

**Step 4**: According to $a_{31} = a_{BW} = 7$, the consistency index (CI) is 3.73, thus, the consistency ratio can be determined $CR = \frac{0.0202}{3.73} = 0.0054$, it is near zero, and this implies a very good consistency.

Therefore, the relative weights of the four groups of stakeholders are 0.0707, 0.1616, 0.5051, and 0.2626, respectively.

According to Eqs. (3.4), (3.9) and (3.10), the weighted direct-influenced matrix can be obtained, as presented in Table 3.2. For instance, the elements of cell (1, 2) in Table 3.3 are N, VL, VL, and VL corresponding to $\Delta^{-1}(s_0, 0) = 0$, $\Delta^{-1}(s_1, 0) = 0.2$, $\Delta^{-1}(s_1, 0) = 0.2$ and $\Delta^{-1}(s_1, 0) = 0.2$ that are provide by the tourist group (group 1), tour guide group (group 2), manager group (group 3), and researcher group (group 4), respectively. Accordingly, the weighted value can be obtained as presented in Eq. (3.31).

$$
\begin{aligned}
\tilde{a}_{12} &= \Delta\left(\sum_{k=1}^{H} \omega_k \Delta^{-1}\left(x_{ij}^k, 0\right)\right) \\
&= \Delta(0.0707 \times 0 + 0.1616 \times 0.2 + 0.5051 \times 0.2 + 0.2626 \times 0.2) = \Delta(0.1859)
\end{aligned}
\tag{3.31}
$$

Accordingly, the element of cell (1,2) in the weighted direct-influenced matrix should be 0.1859. In a similar way, the other elements in the weighted direct-influenced matrix can also be determined (Table 3.4).

**Table 3.4** The weighted direct-influenced matrix by using crisp numbers

| | $B_1$ | $B_2$ | $B_3$ | $B_4$ | $B_5$ | $B_6$ | $B_7$ | $B_8$ | $B_9$ | $B_{10}$ | $B_{11}$ |
|---|---|---|---|---|---|---|---|---|---|---|---|
| $B_1$ | 0.0000 | 0.1859 | 0.1293 | 0.7535 | 0.1010 | 0.5131 | 0.9010 | 0.0000 | 0.9677 | 0.1313 | 0.7313 |
| $B_2$ | 0.1010 | 0.0000 | 0.2990 | 0.3293 | 0.2546 | 0.0000 | 0.4848 | 0.0525 | 0.0465 | 0.3475 | 0.3818 |
| $B_3$ | 0.8626 | 0.8020 | 0.0000 | 0.9010 | 0.4667 | 0.6808 | 0.8222 | 0.5192 | 0.6525 | 0.2869 | 0.5919 |
| $B_4$ | 1.0000 | 0.7818 | 0.0465 | 0.0000 | 0.0000 | 0.4141 | 0.6343 | 0.5535 | 0.8828 | 0.1010 | 0.0000 |
| $B_5$ | 0.2808 | 0.1535 | 0.0000 | 0.0000 | 0.0000 | 0.6000 | 0.9475 | 0.9192 | 0.7616 | 0.0000 | 0.0323 |
| $B_6$ | 0.8687 | 0.0525 | 0.1798 | 0.3475 | 0.1515 | 0.0000 | 0.4162 | 0.4263 | 0.7454 | 0.0323 | 0.2323 |
| $B_7$ | 0.7677 | 0.0000 | 0.0000 | 0.1535 | 0.0000 | 1.0000 | 0.0000 | 0.2283 | 0.9535 | 0.0525 | 0.0465 |
| $B_8$ | 0.1939 | 0.0000 | 0.0000 | 0.0000 | 0.4869 | 0.5677 | 0.6727 | 0.0000 | 0.7535 | 0.0848 | 0.1293 |
| $B_9$ | 0.3313 | 0.6808 | 0.3475 | 0.0323 | 0.0000 | 0.1172 | 0.2869 | 0.0525 | 0.0000 | 0.0525 | 0.1152 |
| $B_{10}$ | 0.9677 | 0.4061 | 0.2283 | 0.1010 | 0.3454 | 0.5798 | 0.5374 | 0.5475 | 0.6990 | 0.0000 | 0.0141 |
| $B_{11}$ | 1.0000 | 0.3010 | 0.0323 | 0.9475 | 0.4546 | 0.6948 | 0.5152 | 0.0525 | 0.6667 | 0.0000 | 0.0000 |

According to Eq. (3.19), the biggest value among the sums of each row and each column s can be determined, it is 7.1293. Subsequently, the normalized initial direct-relation matrix $D$ can be obtained, as presented in Table 3.5. Then, the total relation matrix $T$ can be obtained according to Eqs. (3.22) and (3.23), and the results were presented in Table 3.6.

According to Eqs. (3.24) and (3.25) and Table 3.6. The total effects that directly and indirectly exerted by each factor and the total effects that directly and indirectly exerted by each factor can also be determined, as presented in Table 3.7. Finally, the total effects given and received by each factor and the net effect that contributed by each factor to China's tourism industry can be determined, as resented in Fig. 3.1. These eleven factors were divided into three groups according to the values of $r_{B_j} - c_{B_j}$: the cause group, the effect group, and the linkage group. The threshold value for the cause group is 0.35, it means that the barrier belongs to the cause group if the value of $r_{B_j} - c_{B_j}$ with respect to this barrier is greater than 0.35. The threshold value for the cause group is −0.35, it means that the barrier belongs to the effect group if the value of $r_{B_j} - c_{B_j}$ with respect to this barrier is smaller than −0.35. The barrier belongs to the linkage group if the value of $r_{B_j} - c_{B_j}$ with respect to this barrier is between −0.35 and 0.35. Accordingly, there are four barriers in the cause group including lack of policy support ($B_3$), insufficient governmental investment ($B_{11}$), incomplete legal and regulation system ($B_{10}$), and staff quality ($B_5$). There are also four barriers belonging to effect group weak industrial foundation ($B_1$), weak tourism brand ($B_9$), lack of innovations ($B_7$), and lack collaborations among tourism enterprises ($B_6$). The residual three barriers including lack of complete infrastructure ($B_4$), leisure consciousness ($B_2$), and governance and management ability ($B_8$) belong to the linkage group.

According to Eqs. (3.26) and (3.27), the weights and the normalized weights of the eleven barriers can be determined, as presented in Table 3.8. The relative importance of the eleven barriers can also be divided into three groups by setting two threshold values, namely the 'significantly important group', the 'moderately important group', and the 'weakly important group'. The 'significantly important group' consists of four barriers whose weights are greater than 0.1000, including weak industrial foundation ($B_1$), weak tourism brand ($B_9$), lack of innovations ($B_7$), and lack of policy support ($B_3$). The 'moderately important group' consists of four barriers whose weights are greater than 0.8000 but smaller than 0.1000, including lack collaborations among tourism enterprises ($B_6$), lack of complete infrastructure ($B_4$), and insufficient governmental investment ($B_{11}$). The other barriers whose weights are smaller than 0.8000 (i.e. incomplete legal and regulation system ($B_{10}$), leisure consciousness ($B_2$), governance and management ability ($B_8$), and staff quality ($B_5$)), belong to the 'weakly important group'.

Therefore, lack of policy support ($B_3$), insufficient governmental investment ($B_{11}$), incomplete legal and regulation system ($B_{10}$), and staff quality ($B_5$) belonging to the cause group are the origins of the barriers that hinder the development of China's tourism industry, they are the key driving reasons of these barriers if recognizing China's tourism industry as a system. The weak industrial foundation ($B_1$), weak

**Table 3.5** The normalized initial direct-relation matrix D

| | $B_1$ | $B_2$ | $B_3$ | $B_4$ | $B_5$ | $B_6$ | $B_7$ | $B_8$ | $B_9$ | $B_{10}$ | $B_{11}$ |
|---|---|---|---|---|---|---|---|---|---|---|---|
| $B_1$ | 0.0000 | 0.0261 | 0.0181 | 0.1057 | 0.0142 | 0.0720 | 0.1264 | 0.0000 | 0.1357 | 0.0184 | 0.1026 |
| $B_2$ | 0.0142 | 0.0000 | 0.0419 | 0.0462 | 0.0357 | 0.0000 | 0.0680 | 0.0074 | 0.0065 | 0.0487 | 0.0536 |
| $B_3$ | 0.1210 | 0.1125 | 0.0000 | 0.1264 | 0.0655 | 0.0955 | 0.1153 | 0.0728 | 0.0915 | 0.0402 | 0.0830 |
| $B_4$ | 0.1403 | 0.1097 | 0.0065 | 0.0000 | 0.0000 | 0.0581 | 0.0890 | 0.0776 | 0.1238 | 0.0142 | 0.0000 |
| $B_5$ | 0.0394 | 0.0215 | 0.0000 | 0.0000 | 0.0000 | 0.0842 | 0.1329 | 0.1289 | 0.1068 | 0.0000 | 0.0045 |
| $B_6$ | 0.1218 | 0.0074 | 0.0252 | 0.0487 | 0.0213 | 0.0000 | 0.0584 | 0.0598 | 0.1046 | 0.0045 | 0.0326 |
| $B_7$ | 0.1077 | 0.0000 | 0.0000 | 0.0215 | 0.0000 | 0.1403 | 0.0000 | 0.0320 | 0.1337 | 0.0074 | 0.0065 |
| $B_8$ | 0.0272 | 0.0000 | 0.0000 | 0.0000 | 0.0683 | 0.0796 | 0.0944 | 0.0000 | 0.1057 | 0.0119 | 0.0181 |
| $B_9$ | 0.0465 | 0.0955 | 0.0487 | 0.0045 | 0.0000 | 0.0164 | 0.0402 | 0.0074 | 0.0000 | 0.0074 | 0.0162 |
| $B_{10}$ | 0.1357 | 0.0570 | 0.0320 | 0.0142 | 0.0485 | 0.0813 | 0.0754 | 0.0768 | 0.0980 | 0.0000 | 0.0020 |
| $B_{11}$ | 0.1403 | 0.0422 | 0.0045 | 0.1329 | 0.0638 | 0.0961 | 0.0723 | 0.0074 | 0.0935 | 0.0000 | 0.0000 |

**Table 3.6** The total relation matrix T

| | $B_1$ | $B_2$ | $B_3$ | $B_4$ | $B_5$ | $B_6$ | $B_7$ | $B_8$ | $B_9$ | $B_{10}$ | $B_{11}$ |
|---|---|---|---|---|---|---|---|---|---|---|---|
| $B_1$ | 0.1046 | 0.0829 | 0.0416 | 0.1565 | 0.0373 | 0.1468 | 0.2030 | 0.0411 | 0.2400 | 0.0327 | 0.1322 |
| $B_2$ | 0.0752 | 0.0313 | 0.0522 | 0.0780 | 0.0522 | 0.0516 | 0.1185 | 0.0365 | 0.0744 | 0.0569 | 0.0720 |
| $B_3$ | 0.2548 | 0.1825 | 0.0338 | 0.2016 | 0.1031 | 0.2050 | 0.2450 | 0.1335 | 0.2538 | 0.0642 | 0.1371 |
| $B_4$ | 0.2109 | 0.1503 | 0.0320 | 0.0501 | 0.0246 | 0.1237 | 0.1688 | 0.1053 | 0.2190 | 0.0320 | 0.0431 |
| $B_5$ | 0.1043 | 0.0513 | 0.0176 | 0.0318 | 0.0204 | 0.1428 | 0.1906 | 0.1528 | 0.1896 | 0.0108 | 0.0313 |
| $B_6$ | 0.1866 | 0.0504 | 0.0430 | 0.0919 | 0.0408 | 0.0622 | 0.1295 | 0.0868 | 0.1907 | 0.0171 | 0.0657 |
| $B_7$ | 0.1672 | 0.0372 | 0.0203 | 0.0599 | 0.0160 | 0.1807 | 0.0617 | 0.0564 | 0.2060 | 0.0174 | 0.0381 |
| $B_8$ | 0.0838 | 0.0286 | 0.0152 | 0.0274 | 0.0795 | 0.1258 | 0.1428 | 0.0288 | 0.1713 | 0.0191 | 0.0382 |
| $B_9$ | 0.0876 | 0.1168 | 0.0600 | 0.0369 | 0.0156 | 0.0518 | 0.0833 | 0.0254 | 0.0492 | 0.0191 | 0.0400 |
| $B_{10}$ | 0.2137 | 0.1012 | 0.0552 | 0.0666 | 0.0735 | 0.1532 | 0.1638 | 0.1137 | 0.2047 | 0.0168 | 0.0457 |
| $B_{11}$ | 0.2328 | 0.0979 | 0.0294 | 0.1845 | 0.0834 | 0.1701 | 0.1672 | 0.0540 | 0.2117 | 0.0171 | 0.0430 |

**Table 3.7** The total effects that directly and indirectly exerted by each factor and the total effects that directly and indirectly exerted by each factor

| | $B_1$ | $B_2$ | $B_3$ | $B_4$ | $B_5$ | $B_6$ | $B_7$ | $B_8$ | $B_9$ | $B_{10}$ | $B_{11}$ |
|---|---|---|---|---|---|---|---|---|---|---|---|
| $r_{B_j}$ | 1.2186 | 0.6988 | 1.8144 | 1.1599 | 0.9434 | 0.9648 | 0.8607 | 0.7605 | 0.5857 | 1.2980 | 1.2911 |
| $c_{B_j}$ | 1.7214 | 0.9302 | 0.4002 | 0.9852 | 0.5464 | 1.4137 | 1.6741 | 0.8343 | 2.0105 | 0.3033 | 0.6864 |

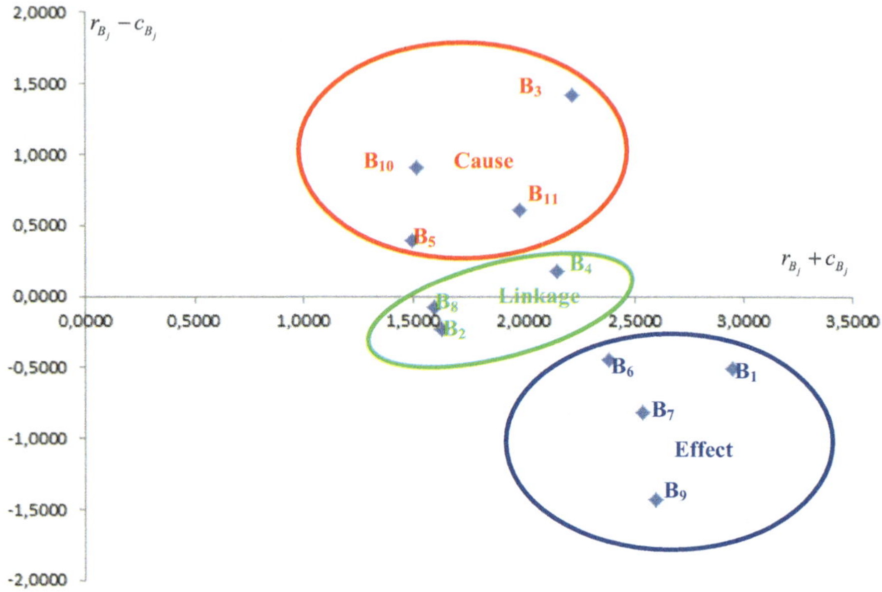


**Fig. 3.1** The cause-effect diagram of the barriers in China's tourism industry

tourism brand ($B_9$), lack of innovations ($B_7$), and lack of policy support ($B_3$) belonging to the 'significantly important group' are the important barriers that hinder the development of China's tourism industry. It is apparent that lack of policy support ($B_3$) belongs to both the cause group and the 'significantly important group', thus, this barrier-'lack of policy support' is the most important origin that hinders the development of China's tourism industry. The barrier 'insufficient governmental investment ($B_{11}$)' is the secondly most important origin that hinders the development of China's tourism industry. Incomplete legal and regulation system ($B_{10}$) and staff quality ($B_5$) are two barriers belonging to the cause group; however, the relative importance of both the two barriers is low, and they are raked at the eighth and eleventh (the last) position.

The four barriers including industrial foundation ($B_1$), weak tourism brand ($B_9$), lack of innovations ($B_7$), and lack of policy support ($B_3$) are the four most important barriers. However, it is worth pointing out that the industrial foundation ($B_1$), weak tourism brand ($B_9$), lack of innovations ($B_7$) belong to the effect group, it means that these three barriers are not the origins of the problem existed in China's tourism industry. The lack collaborations among tourism enterprises ($B_6$) belonging to the effect group was also ranked at the fifth position. The barrier-lack of complete infrastructure ($B_4$) belonging to the linkage group was ranked at the sixth position.

**Table 3.8** The weights and the normalized weights of the eleven barriers

| | $B_1$ | $B_2$ | $B_3$ | $B_4$ | $B_5$ | $B_6$ | $B_7$ | $B_8$ | $B_9$ | $B_{10}$ | $B_{11}$ |
|---|---|---|---|---|---|---|---|---|---|---|---|
| $\bar{\omega}_j$ | 2.9828 | 1.6453 | 2.6276 | 2.1522 | 1.5417 | 2.4204 | 2.6621 | 1.5966 | 2.9615 | 1.7614 | 2.0678 |
| $\omega_j$ | 0.1221 | 0.0674 | 0.1076 | 0.0881 | 0.0631 | 0.0991 | 0.1090 | 0.0654 | 0.1213 | 0.0721 | 0.0847 |
| Ranking | 1 | 9 | 4 | 6 | 11 | 5 | 3 | 10 | 2 | 8 | 7 |

## 3.5 Policy Implications and Conclusions

According to the obtained results in Sect. 3.4, the following policy implications can be obtained for China's decision-makers of China's tourism industry

(1) Various policy support measures are urgently needed to China's tourism industry. China's administration should establish various policy support measures to promote the further development China's tourism industry, i.e. low interest or zero interest loans, subsidies, and some other policy measures for guaranteeing the healthy and harmonious development of China's tourism industry.
(2) Sufficient investment from the government is prerequisite to China's tourism industry. China's government should invest more on the infrastructure to promote the further development of the tourism industry.
(3) China's administration should draft more regulations, laws and legislations to establish complete legal and regulation system for guarantee all the stakeholders in value chain of the tourism industry.
(4) Staff quality also plays an important role for addressing the barriers existed in China's tourism industry, thus, the improvement of staff quality is of vital importance. Accordingly, more strict professional qualifications should be set for people to enter tourism industry, and the training and education on the tourism staff should also be carried out to improve staff quality.
(5) The most severe weak points of China's tourism industry are weak industrial foundation, weak tourism brand, and lack of innovations, thus, China's administration should take various effective strategic measures to enhance the industrial foundation of China's tourism industry, improve the tourism brand, and encourage innovations tourism industry.

All in all, this study aims at studying the barriers that hinder the development of China's tourism industry and providing effective measures to promote the development of China's tourism industry. The critical barriers were firstly summarized, then, an improved DEMATEL method was developed to investigate the cause-effect relationships among these barriers and prioritize the barriers according to their relative importance. Ten barriers including weak industrial foundation, leisure consciousness, lack of policy support, lack of complete infrastructure, staff quality, lack collaborations among tourism enterprises, lack of innovations, governance and management ability, weak tourism brand, incomplete legal and regulation system, and insufficient governmental investment, were studied in this paper, and the 2-tuple DEMATEL was employed to analyze the complex relationships among these barriers and rank these barriers, and some policy implications were presented to China's decision-makers for them to promote the development of China's tourism industry.

## References


G.I. Alptekin, G. Büyüközkan, An integrated case-based reasoning and MCDM system for Web based tourism destination planning. Expert Syst. Appl. **38**(3), 2125–2132 (2011)

A. Baykasoğlu, V. KaplanoğLu, Z.D. DurmuşOğLu, C. ŞAhin, Integrating fuzzy DEMATEL and fuzzy hierarchical TOPSIS methods for truck selection. Expert Syst. Appl. **40**(3), 899–907 (2013)

J. Bao, G. Chen, L. Ma, Tourism research in China: insights from insiders. Ann. Tour. Res. **45**, 167–181 (2014)

C.A. Chen, Using DEMATEL method for medical tourism development in Taiwan. Am. J. Tour. Res. **1**(1), 26–32 (2012)

S. Chen, L. Ni, Analysis on obstacles to the tourist industry in China. J. Xi'an Jiaotong Univ. (Soc. Sci.) **21**(4), 62–64 (2001)

C.T. Chen, W.S. Tai, Measuring the intellectual capital performance based on 2-tuple fuzzy linguistic information, in: *The 10th Annual Meeting of APDSI, Asia Pacific Region of Decision Sciences Institute*, vol. 20 (2005)

F.H. Chen, T.S. Hsu, G.H. Tzeng, A balanced scorecard approach to establish a performance evaluation and relationship model for hot spring hotels based on a hybrid MCDM model combining DEMATEL and ANP. Int. J. Hosp. Manag. **30**(4), 908–932 (2011)

D. Dalalah, M. Hayajneh, F. Batieha, A fuzzy multi-criteria decision making model for supplier selection. Expert Syst. Appl. **38**(7), 8384–8391 (2011)

L. Dwyer, P. Forsyth, R. Spurr, S. Hoque, Economic impacts of a carbon tax on the Australian tourism industry. J Travel Res. **52**(2), 143–155 (2013)

E. Fontela, A. Gabus, The DEMATEL observer. In: DEMATEL 1976 Report Switzerland, Geneva: Battelle Geneva Research Center (1976)

A. Gabus, E. Fontela, Perceptions of the world problematique: communication procedure, communicating with those bearing collective responsibility. DEMATEL Report No. 1. Geneva, Switzerland: Battelle Geneva Research Center (1973)

H. Liang, J. Ren, Z. Gao, S. Gao, X. Luo, L. Dong, A. Scipioni, Identification of critical success factors for sustainable development of biofuel industry in China based on grey decision-making trial and evaluation laboratory (DEMATEL). J Clean Prod. (2016) (in press)

S.S. Hassan, Determinants of market competitiveness in an environmentally sustainable tourism industry. J. Travel Res. **38**(3), 239–245 (2000)

F. Herrera, L. Martínez, A 2-tuple fuzzy linguistic representation model for computing with words. IEEE Trans. Fuzzy Syst. **8**(6), 746–752 (2000)

W.R.J. Ho, C.L. Tsai, G.H. Tzeng, S.K. Fang, Combined DEMATEL technique with a novel MCDM model for exploring portfolio selection based on CAPM. Expert Syst. Appl. **38**(1), 16–25 (2011)

H.Y. Hu, S.I. Chiu, C.C. Cheng, T.M. Yen, Applying the IPA and DEMATEL models to improve the order-winner criteria: a case study of Taiwan's network communication equipment manufacturing industry. Expert Syst. Appl. **38**(8), 9674–9683 (2011)

Y. Jeon, J. Kim, An application of SWOT-AHP to develop a strategic planning for a tourist destination (2011)

Ö. Kabak, F. Ülengin, B. Çekyay, Ş. Önsel, Ö. Özaydın, Critical success factors for the iron and steel industry in turkey: a fuzzy DEMATEL approach. Int. J. Fuzzy Syst. 1–14 (2015)

D. Leung, G. Li, L.H.N. Fong, R. Law, A. Lo, Current state of China tourism research. Curr. Issues Tour. **17**(8), 679–704 (2014)

Li, Problem in the development of China's tourism industry and countermeasures. Guangming Daily. (2013) http://epaper.gmw.cn/gmrb/html/2013-09/22/nw.D110000gmrb_20130922_4-07.htm

C.H. Liu, G.H. Tzeng, M.H. Lee, Improving tourism policy implementation—The use of hybrid MCDM models. Tour. Manag. **33**(2), 413–426 (2012)

C.H. Liu, G.H. Tzeng, M.H. Lee, P.Y. Lee, Improving metro–airport connection service for tourism development: using hybrid MCDM models. Tour Manag. Perspect **6**, 95–107 (2013)

H.C. Liu, J.X. You, C. Lu, Y.Z. Chen, Evaluating health-care waste treatment technologies using a hybrid multi-criteria decision making model. Renew. Sust. Energ. Rev. **41**, 932–942 (2015)

C. Morrison, C.M. Pickering, Perceptions of climate change impacts, adaptation and limits to adaption in the Australian Alps: the ski-tourism industry and key stakeholders. J. Sust. Tour. **21**(2), 173–191 (2013)

J. Rezaei, Best-worst multi-criteria decision-making method. Omega **53**, 49–57 (2015a)

J. Rezaei, (2015b) Best-worst multi-criteria decision-making method: some properties and a linear model. Omega (2015b) (in press)

J. Ren, B.K. Sovacool, Quantifying, measuring, and strategizing energy security: determining the most meaningful dimensions and metrics. Energy **76**, 838–849 (2014)

J. Ren, A. Manzardo, S. Toniolo, A. Scipioni, Sustainability of hydrogen supply chain. Part I: identification of critical criteria and cause-effect analysis for enhancing the sustainability using DEMATEL. Int. J. Hydrog. Energy **38**(33), 14159–14171 (2013)

T.L. Saaty, *The Analytic Hierarchy Process* (McGraw Hill, New York, 1980)

Z. Tang, J. Shang, C. Shi, Z. Liu, K. Bi, Decoupling indicators of $CO_2$ emissions from the tourism industry in China: 1990–2012. Ecol. Indic. **46**, 390–397 (2014)

W.H. Tsai, J.L. Hsu, C.H. Chen, W.R. Lin, S.P. Chen, An integrated approach for selecting corporate social responsibility programs and costs evaluation in the international tourist hotel. Int. J. Hosp. Manag. **29**(3), 385–396 (2010)

S.H. Tsaur, C.H. Wang, The evaluation of sustainable tourism development by analytic hierarchy process and fuzzy set theory: an empirical study on the Green Island in Taiwan. Asia Pac. J. Tour. Res. **12**(2), 127–145 (2007)

M.L. Tseng, Using the extension of DEMATEL to integrate hotel service quality perceptions into a cause-effect model in uncertainty. Expert Syst. Appl. **36**(5), 9015–9023 (2009)

Y. Yuan, W. Li, Stratify obstacles of inter-regional tourism industry chain based on diagnosis model: Shaoshan, Huaminglou, Wushi as an example. J. Beijing Int. Stud. Univ. **199**, 29–33 (2011)

H. Zhang, C.L. Gu, L.W. Gu, Y. Zhang, The evaluation of tourism destination competitiveness by TOPSIS & information entropy—A case in the Yangtze river Delta of China. Tour. Manag. **32**(2), 443–451 (2011)

P. Zhou, The obstacles to the peri-urban leisure travel industry development and its correction: Chongqing case. Chongqing Soc. Sci. **179**, 41–45 (2009). (in Chinese)

# Chapter 4
# Fuzzy Best-Worst Method and Interpretive Structural Modelling for Complex System Analysis: Enablers Analysis for Aviation Maintenance Safety

**Abstract** This study aims at employing the multi-criteria decision analysis method for analyzing the enablers of aviation maintenance safety, the twelve enablers of aviation maintenance safety in human, facilities, institution and management aspects were firstly summarized; subsequently, the fuzzy best-worst network method was used to prioritize these enablers according to their relative importance, and the significantly important and moderately important enablers can be identified; then, eight strategic measures were proposed according to the relative importance of the enablers; finally, the interpretive structural modelling was employed to investigate the complex relationships among these eight strategic measures, and drafting appropriate plan and schedule, training on aviation maintenance, education on safety maintenance awareness, perfect the regulation and standard system, and establishing complete safety management system should be adopted by the decision-makers/stakeholders to improve the aviation maintenance safety.

## 4.1 Introduction

The implementation of safety culture and safety management can effectively reduce accident rate and prevent and resolve the safety incidents (Xia et al. 2010). Aviation maintenance safety has attracted more and more attentions, because the errors in aviation maintenance may lead to serious aviation accidents. It was reported that 12% of the total accidents were caused by maintenance and inspection faults, and the human errors may occur within the maintenance arena (Endsley and Robertson 2000). The objective of aviation maintenance is to provide the serviceable aircraft when it is required by the operator at minimum cost (Knotts 1999). The successful aviation maintenance needs not only the skilled maintenance personnel but also maintenance facilities and safety management system, However, there are various factors influencing aviation maintenance safety such as close working cooperation and coordination among the job skills (Kim and Song 2015), facilities advancement and completion, regulations and standards, and supply of aircraft materials, etc. And there are also various interdependent and interacted relationships among these influential factors, meanwhile the relative importance of these influential factors



J. Ren, *Advanced Operations Management for Complex Systems Analysis*,
SpringerBriefs in Applied Sciences and Technology,
https://doi.org/10.1007/978-3-030-45418-0_4

differs. In order to draft effective measures for improving aviation maintenance safety, the prioritization of these influential factors is prerequisite.

After determining the strategic measures for improving aviation maintenance safety, it is also difficult for the decision-makers/stakeholders to identify the complex relationships among these strategic measures. This study employed the fuzzy best-worst network method which can consider the interacted relationships among the influential factors was employed to determine the relative weights of the influential factors, and the interpretive structural modeling method was used to identify the complex relationships among the strategic measures. Based on these, the implications and the strategic measures can be obtained for improving the aviation maintenance safety.

Besides the Introduction part, the residual parts of this study has been organized as follows: the fuzzy best-worst network method and the interpretive structural modeling method were presented in Sect. 4.2; the results were presented in Sect. 4.3; and this study has been concluded in Sect. 4.4.

## 4.2 Methods

In this section, the enablers of aviation maintenance safety were firstly summarized; then, the fuzzy best-worst method which can be used for weights determination without the considerations of the interdependences and interactions among the criteria and the fuzzy best-worst network method which can consider the interdependences and interactions among the criteria were employed to determine the relative importance of the enablers of aviation maintenance safety; finally, the interpretative structural modeling was used to analyze the strategic measures for improving aviation maintenance safety.

### *4.2.1 Enablers of Aviation Maintenance Safety*

A total of twelve enablers in four categories including human aspect, facilities aspect, institution aspect, and management aspect were obtained in this study based on literature reviews and focus group meeting, and they have been specified as follows:

(1) Human aspect (H)

- Safety maintenance style and awareness ($H_1$): safety maintenance style refers to the strong safety perceptions, high compliance with maintenance rules and regulations, and rigorous working style in maintenance personnel (Xia et al. 2010).
- Excellent collaborations and communication ($H_2$): the excellent collaborations and communication of the maintenance personnel with the crew are the

foundation of good understanding of the unit failure, and then the maintenance personnel can appropriate mend the equipment to guarantee the aviation safety (Zhang et al. 2016).
- Excellent and mature operation skills ($H_3$): the excellent basic skills and mature operations of the maintenance personnel can effectively ensure the maintenance working quality and the maintenance efficiency (Gao and Wang. 2016).

(2) Facilities aspect (F)

- Appropriate use of maintenance facilities ($F_1$): the safe keeping, use and check of the maintenance facilities is critical for ensure the maintenance accuracy, and these will further influence the aviation safety (Gao and Wang 2016).
- Completion of factories and equipment ($F_2$): the completion of factories and equipment is the key for assuring the maintenance efficiency and quality. In addition, it also has significant effects on the safety of the maintenance personnel (Gao and Wang 2016).
- Sufficient supply of aircraft materials ($F_3$): The insufficient supply of aircraft materials will influence the development of maintenance, and the maintenance personnel cannot successfully finish the maintenance task (Gao and Wang 2016).

(3) Institution aspect (I)

- Regulations and standards ($I_1$): the inappropriate and unscientific regulations, policies and standards will make the maintenance personnel difficult to follow, and this may lead to unsafe or maintenance failure events (Gao et al. 2010).
- Appropriate engineering and technological documents ($I_2$): the engineering and technological aviation maintenance plan can effectively provide the technological support to guarantee aviation maintenance safety and reduce maintenance costs (Gao and Wang 2016).
- Working environment ($I_3$): the physical conditions of the working environment (i.e. temperature, noise, lighting, and humidity, etc.) can influence not only the technical quality of aircraft and the accuracy of the detecting instruments, but also the mood of the maintenance personnel.

(4) Management aspect (M)

- Safety management system ($M_1$): the complete and perfect safety management system will make the maintenance personnel treat the maintenance carefully and rigorously to avoid the mistakes and errors during maintenance (Gao and Wang 2016).
- Appropriate plan and schedule ($M_2$): the appropriate plan and schedule of maintenance work means the appropriate division of labour and tasks, and

suitably coordinate the relationships among human-machine-environment (Gao and Wang 2016).

- Complete labor training system ($M_3$): establishing the complete labor training system can effectively improve the maintenance quality of maintenance personnel, and a team of skilled people is the key to reduce maintenance failure (Gao and Wang 2016).

## 4.2.2 *Fuzzy Best-Worst Method*

Comparing with the traditional AHP and various modified AHP methods, the BWM developed by Rezaei (2015) as a power tool which is more convenient for users to achieve overall consistency when using it for weights determination. As for a problem with n elements, the traditional AHP and various modified AHP methods need to carry out $n(n-1)/2$ times of comparisons; however, the BW method only needs $(2n-3)$ times of comparisons, and it is obvious that when $n > 3$, $(2n-3) < n(n-1)/2$. Accordingly, the BWM has been widely used. For instance, Ren et al. (2017) used the BWM to prioritize the urban sewage sludge treatment technologies. Gupta and Barua (2017) employed the BWM for supplier selection. Abadi et al. (2018) used the BWM to evaluate the medical tourism development. Hafezalkotob and Hafezalkotob (2017) extended the best-worst method to fuzzy environmental for using triangular numbers to reflect the inherent ambiguity of the judgments of the decision-makers, and the fuzzy best-worst method (FBWM) consists of five steps (Rezaei 2015; Hafezalkotob and Hafezalkotob 2017):

**Step 1**: Determining the decision criteria. All the decision and evaluation criteria $\{C_1, C_2, \ldots, C_n\}$ will be determined in this step. As for the evaluation of the aviation maintenance safety, the evaluation criteria may consists of safety perception of maintenance personnel, knowledge and experience, training, safety equipment, communication, and policy and regulation, etc.

**Step 2**: Determining the best (e.g. the most favorable, most important, and most significant) and the worst (e.g. the least favorable, least important, and least significant) criteria among the n criteria, denotes by $C_B$ and $C_W$, respectively. In this step, the decision-makers will determine the best and the worst criteria as references for comparing with other criteria.

**Step 3**: Determining the Best-to-Others (BO) vector by comparing the relative importance/priority of the best criterion with all the criteria. In this step, the decision-makers should specify the preference degree of the best criterion $C_B$ over each of the criteria by using the linguistic terms presented in Table 4.1. And all the linguistic terms can be transformed into triangular fuzzy numbers. Accordingly, the fuzzy BO vector can be determined, as presented in Eq. (4.1):

**Table 4.1** Fuzzy scales used in FBWM for establishing the BO and OW vectors (modified from Guo and Zhao (2017))

| Linguistic terms | Abbreviation | Fuzzy numbers |
|---|---|---|
| Very importance | V | (7/2, 4, 9/2) |
| Adjacent F and V | AFV | (3, 7/2, 4) |
| Fairly importance | F | (5/2, 3, 7/2) |
| Adjacent M and F | AMF | (2, 5/2, 3) |
| Moderate importance | M | (3/2, 2, 5/2) |
| Adjacent W and M | AWM | (1, 3/2, 2) |
| Weakly importance | W | (2/3, 1, 3/2) |
| Completely equal importance | C | (1, 1, 1) |

$$BO = \left[\tilde{a}_{B1}\ \tilde{a}_{B2}\ \ldots\ \tilde{a}_{Bn}\right] \tag{4.1}$$

where *BO* represents the Best-to-Other vector, and $\tilde{a}_{Bj}(j = 1,\ 2,\ \ldots,\ n)$ represents the preference of the best criterion $C_B$ over the *j*-th criterion.

**Step 4**: Determining the Others-to-Worst (OW) vector by comparing the relative importance/priority of all the criteria with the worst criterion. In this step, the decision-makers should specify the preference degree of each of the criteria over the worst criterion $C_W$ by using the linguistic terms presented in Table 4.1. And all the linguistic terms can also be transformed into triangular fuzzy numbers. Accordingly, the fuzzy BW vector can be determined, as presented in Eq. (4.2):

$$OW = \left[\tilde{a}_{1W}\ \tilde{a}_{2W}\ \ldots\ \tilde{a}_{nW}\right] \tag{4.2}$$

where *OW* represents the Other-to-Worst vector, and $\tilde{a}_{jW}(j = 1,\ 2,\ \ldots,\ n)$ represents the preference of the *j*-th criterion over the worst criterion $C_W$.

**Step 5**: Determining the optimum weights of the n criteria $\left[\omega_1^*\ \omega_2^*\ \cdots\ \omega_n^*\right]$.

According to the BO and OW vectors, the weights of the n criteria should satisfy that $\frac{\omega_B}{\omega_j} = \tilde{a}_{Bj}(j = 1,\ 2,\ \ldots,\ n)$ and $\frac{\omega_j}{\omega_W} = \tilde{a}_{jW}(j = 1, 2, \ldots, n)$, and these can be achieved by minimizing the maximum absolute difference $\left|\frac{\omega_B}{\omega_j} - \tilde{a}_{Bj}\right|$ and $\left|\frac{\omega_j}{\omega_W} - \tilde{a}_{jW}\right|$ for all *j*. Accordingly, the constrained optimization programming for determining the optimum weights can be obtained with the consideration of weights normalization:

$$\begin{aligned} &\min \max_j \left\{\left|\frac{\omega_B}{\omega_j} - \tilde{a}_{Bj}\right|, \left|\frac{\omega_j}{\omega_W} - \tilde{a}_{jW}\right|\right\} \\ &s.t. \\ &\sum_{j=1}^{n} \omega_j = 1 \\ &\omega_j \geq 0, j = 1, 2, \ldots, n \end{aligned} \tag{4.3}$$

where $\omega_B$, $\omega_W$, and $\omega_j$ represent the weights of the best criterion, the worst criterion, and the $j$-th criterion, respectively.

The programming (4.3) can be transformed into the following linear programming model according to Rezaei (2015):

$$\begin{aligned}
&\min \max_j \left\{ \left| \omega_B - \tilde{a}_{Bj}\omega_j \right|, \left| \omega_j - \tilde{a}_{jW}\omega_W \right| \right\} \\
&s.t. \\
&\sum_{j=1}^{n} \omega_j = 1 \\
&\omega_j \geq 0, j = 1, 2, \ldots, n
\end{aligned} \tag{4.4}$$

The programming (4.4) can be further transformed into programming (4.5):

$$\begin{aligned}
&\min \xi \\
&s.t. \\
&\left| \omega_B - \tilde{a}_{Bj}\omega_j \right| \leq \xi \\
&\left| \omega_j - \tilde{a}_{jW}\omega_W \right| \leq \xi \\
&\sum_{j=1}^{n} \omega_j = 1 \\
&\omega_j \geq 0, j = 1, 2, \ldots, n
\end{aligned} \tag{4.5}$$

The programming (4.5) is equivalent to programming (4.6):

$$\begin{aligned}
&\min \xi \\
&s.t. \\
&\omega_B - \xi \leq \tilde{a}_{Bj}\omega_j \\
&\omega_B + \xi \geq \tilde{a}_{Bj}\omega_j \\
&\omega_j - \xi \leq \tilde{a}_{jW}\omega_W \\
&\omega_j + \xi \geq \tilde{a}_{jW}\omega_W \\
&\sum_{j=1}^{n} \omega_j = 1 \\
&\omega_j \geq 0, j = 1, 2, \ldots, n
\end{aligned} \tag{4.6}$$

The $\tilde{a}_{Bj}$ and $\tilde{a}_{jW}\left(j = 1, 2, \ldots, n\right)$ in programming (4.6) are fuzzy numbers, and they crisp equivalents can be obtained through the predefined possibility, and then the programming (4.6) can be further transformed into programming (4.7) (Hafezalkotob and Hafezalkotob 2017):

$$
\begin{aligned}
&\min \xi \\
&s.t. \\
&\omega_B - \xi \le \left[a_{Bj}^M + (1-\alpha)a_{Bj}^U\right]\omega_j \\
&\omega_B + \xi \ge \left[a_{Bj}^M - (1-\alpha)a_{Bj}^L\right]\omega_j \\
&\omega_j - \xi \le \left[a_{jW}^M + (1-\alpha)a_{jW}^U\right]\omega_W \\
&\omega_j + \xi \ge \left[a_{jW}^M - (1-\alpha)a_{jW}^L\right]\omega_W \\
&\sum_{j=1}^{n} \omega_j = 1 \\
&\omega_j \ge 0, j = 1, 2, \ldots, n
\end{aligned} \tag{4.7}
$$

where $\alpha$ represents the predefined possibility level determined by the decision-makers.

After solving programming (4.7) for a given predefined possibility level, the minimum value of the objective function $\xi^*$ and the optimum weights of the $n$ weights $\left[\omega_1^* \ \omega_2^* \ \ldots \ \omega_n^*\right]$ can be obtained. $\xi^*$ represents the overall consistency level of the comparisons, and the smaller the value, the more consistent the comparisons are.

### 4.2.3 *Fuzzy Best-Worst Network Method*

The FSWM can effectively and conveniently determine the weights of the criteria, but it cannot incorporate the interdependences and interactions among these criteria, thus, a Fuzzy Best-Worst Network Method (FBWNM) was developed for determining the weights of the criteria with the considerations of the interdependences and interactions among these criteria. Assuming that there are a total of $P$ ($p = 1, 2, \ldots, P$) categories and there are $N_p$ criteria in the $p$-th category, and there are four steps in the developed (FBWNM) including (i) determining the independent weights of the P categories as well as that of the criteria in each category; (ii) determining the inner dependency matrix; (iii) calculating the normalized inter-dependent weights of the $P$ categories; and (iv) determining the final inter-dependent weights of the criteria. These four steps were specified as follows based on the works of Dağdeviren et al. (2008) and Zamani et al. (2014):

**Step 1**: Determining the independent weights of the P categories as well as that of the criteria in each category. In this study, the FBWM was employed to determine the weights of the $P$ categories and that of the criteria in each category under the assumption that all the categories are independent. In other words, the weights of the $P$ categories and that of the criteria in each category were determined without considering the interdependences and interactions among them.

$$W_{\text{category}} = [w_1, w_2, \ldots, w_P] \tag{4.8}$$

where $W_{\text{category}}$ represents the weight vector of the $P$ categories, and $w_t (t = 1, 2, \ldots, P)$ represents the weights of the $t$-th category.

**Step 2**: Determining the inner dependency matrix. The inner dependency matrix $D$ of the $P$ categories with respect to each category can also be determined by FBWM. The elements of the $t$-th column vector in matrix $D$ represent the relative effects of the categories on the $t$-th category, and this vector can be obtained through establishing the comparison matrix with respect to the $t$-th category. In a similar way, all the column vectors in matrix $D$ can be obtained.

$$D = \begin{vmatrix} 1 & d_{12} & \cdots & d_{1P} \\ d_{21} & 1 & \cdots & d_{2P} \\ \vdots & \cdots & \ddots & \vdots \\ d_{P1} & d_{P2} & \cdots & 1 \end{vmatrix} \tag{4.9}$$

where $D$ is the inner dependency matrix, and $d_{pt}$ represents the relative effect of the $p$-th category on the $t$-th category.

**Step 3**: Calculating the normalized inter-dependent weights of the $P$ categories. The inter-dependent weights of the $P$ categories can be determined by Eq. (4.10), then the normalized inter-dependent weights of the $P$ categories can be determined by Eq. (4.11).

$$W' = D \times W_1^T = [\omega_1, \omega_2, \ldots, \omega_n] \tag{4.10}$$

$$W = \left[ \omega_1 / \sum_{i=1}^{n} \omega_i, \omega_2 / \sum_{i=1}^{n} \omega_i, \ldots, \omega_n / \sum_{i=1}^{n} \omega_i \right] \tag{4.11}$$

where $W'$ represents weight vector of the inter-dependent priorities of the $P$ categories, and $W$ represents the normalized weight vector of the inter-dependent weights of the $P$ categories.

**Step 4**: Determining the final inter-dependent weights of the criteria. The final inter-dependent weights of each criterion can be determined by calculating the product of the independent weight of the criterion and the normalized inter-dependent weight of the corresponding category.

### *4.2.4 Interpretative Structural Modeling*

Interpretative Structural Modeling (ISM) developed by Warfield (1974a, b) is a computer-assistant tool for studying the interdependent and interacted relationships among the elements of a complex system (Ren et al. 2015). In this study, the ISM method was used to investigate the complex cause-effect relationships among the strategic measures for improving the aviation maintenance safety which were recommended based on the results of the Fuzzy Best-Worst Network Method. The procedures of ISM were illustrated in Fig. 4.1, and the procedures were specified as follows (Warfield 1974a, b; Agi and Nishant 2017).

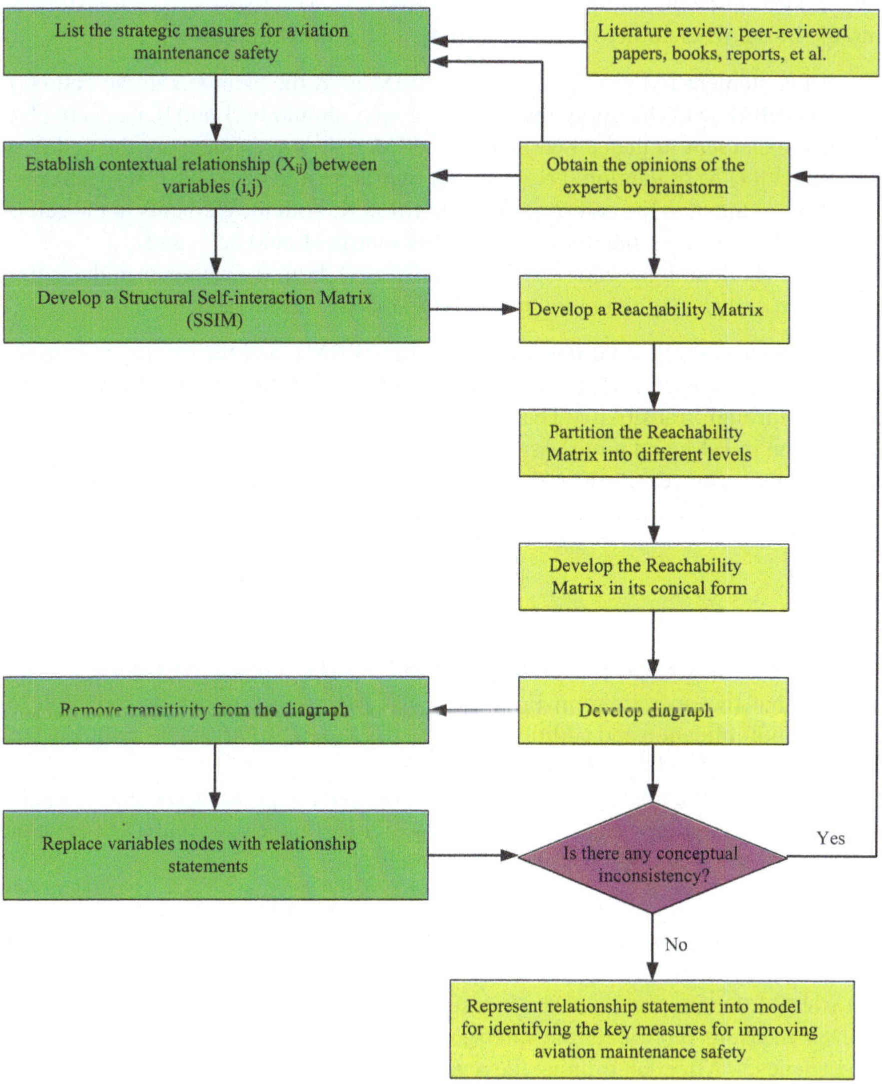


**Fig. 4.1** Procedures of ISM methodology for identifying the measure for aviation maintenance (modified from Kannan et al. 2009)

**Step 1**: Determine the structural self-interaction matrix (SSIM). 'V', 'A', 'X', and 'O' can be used to represent the relationship a strategic measure over another.

V: The $i$-th measure will exert the $j$-th measure;
A: The $j$-th measure will be exerted by the $i$-th measure;
X: The $i$-th measure and the $j$-th measure will exert each other; and
O: No direct relationship between the $i$-th measure and the $j$-th measure.

**Step 2**: Determine the initial reachability matrix (A). The SSIM will be transformed into the initial reachability matrix according to the following rules:

(1) If the element in the cell $(I, j)$ of the SSIM is V, the elements in the cell $(I, j)$ and cell $(j, i)$ of the initial reachability matrix should be 1 and 0, respectively;
(2) If the element in the cell $(I, j)$ of the SSIM is A, the elements in the cell $(I, j)$ and cell $(j, i)$ of the initial reachability matrix should be 0 and 1, respectively;
(3) If the element in the cell $(I, j)$ of the SSIM is X, both the elements in the cell $(I, j)$ and cell $(j, i)$ of the initial reachability matrix should be 1; and
(4) If the element in the cell $(I, j)$ of the SSIM is O, both the elements in the cell $(I, j)$ and cell $(j, i)$ of the initial reachability matrix should be 0.

**Step 3**: Determine the final reachability matrix (FRM), and the driving power and dependence power of each measure. The final reachability matrix (FRM) can be obtained through transitivity. The transitivity of the contextual relation is a basic assumption in ISM, and it represents that if the $i$-th measure exerts the $j$-th measure, and the $j$-th measure exerts the $k$-th measure, then the $i$-th measure is regarded to exert the $k$-th measure. Mathematically, the final reachability matrix can be determined by Eq. (4.12):

$$R = (A + I)^{r-1} = (A + I)^{r} \tag{4.12}$$

where $r$ is an integer which is greater to make $(A + I)^{r-1}$ equal to $(A + I)^{r}$.

Note that the operations in Eq. (4.12) was carried out under the principle of Boolean multiplication and addition ($1 * 1 = 1$, $1 + 1 = 1$, $1 * 0 = 0 * 1 = 0$, $1 + 0 = 0 + 1 = 1$, $0 * 0 = 0$ and $0 + 0 = 0$).

According to the final reachability matrix, the driving power and the dependence power of each measure can be determined by all ones in the rows and all ones on the columns with respect that measure, respectively.

**Step 4**: Level partitions. The reachability set with respect to each strategic measure consists of all the measures which were exerted by that measure and the measure itself. The antecedent set with respect to each strategic measure consists of all the measure that exerted that measure and the measure itself. The intersection set with respect to each measure represents the intersection between the reachability set and the antecedent set. After determining the reachability set, antecedent set, and interaction set with respect to each measure from the final reachability matrix, the measures belonging to level I in ISM hierarchy can be determined through identifying the measures with respect to which the reachability set and intersection set are the same.

The measures belonging to level I should be discarded to identify the measures belonging to level II in ISM hierarchy through repeating the process of finding the measures belonging to level I. Similarly, the measures in the other levels of ISM hierarchy can be identified with iterations.

**Step 5**: Determine the ISM hierarchy model. According to the final reachability matrix, and various ISM hierarchy levels determined in Step 5, the measures in the same level and that in the adjacent levels are connected in the format of diagraph by vertices and edges. Then, the ISM hierarchy model can be determined after removing the transitivities among the measures.

## 4.3 Results and Discussion

The fuzzy best-worst network method was employed to determine the normalized inter-dependent weights of the four aspects of aviation maintenance safety, and the fuzzy best-worst method was firstly used to determine independent weights of the four categories. Human aspect and facilities aspect were recognized as the most sustainable and the least sustainable category, respectively (Table 4.2).

The BW vector and the OW vector can be firstly determined through compare the relative importance of the human aspect with all the four aspects as well as that of the four aspects comparing with facilities aspect by using linguistic terms, and all the linguistic terms can be transformed into triangular numbers. Then, the BO and OW vectors for determining the independent weights of the four categories by triangular fuzzy numbers can be determined, as presented in Table 4.3.

According to Table 4.3, the programming for determining the weights of the four categories can be determined:

**Table 4.2** The BO and OW vectors for determining the independent weights of the four categories by using linguistic terms

| | Human aspect (H) | Facilities aspect (F) | Institution aspect (I) | Management aspect (M) |
|---|---|---|---|---|
| BO | C | F | M | W |
| OW | F | C | M | F |

**Table 4.3** The BO and OW vectors for determining the independent weights of the four categories by using triangular fuzzy numbers

| | Human aspect (H) | Facilities aspect (F) | Institution aspect (I) | Management aspect (M) |
|---|---|---|---|---|
| BO | (1, 1, 1) | (5/2, 3, 7/2) | (3/2, 2, 5/2) | (2/3, 1, 3/2) |
| OW | (5/2, 3, 7/2) | (1, 1, 1) | (3/2, 2, 5/2) | (5/2, 2, 7/2) |

$$
\begin{aligned}
&\min \xi \\
&s.t. \\
&\omega_H - \xi \le \left[a_{HF}^M + (1-\alpha)a_{HF}^U\right]\omega_F \\
&\omega_H + \xi \ge \left[a_{HF}^M - (1-\alpha)a_{HF}^L\right]\omega_F \\
&\omega_H - \xi \le \left[a_{HI}^M + (1-\alpha)a_{HI}^U\right]\omega_I \\
&\omega_H + \xi \ge \left[a_{HI}^M - (1-\alpha)a_{HI}^L\right]\omega_I \\
&\omega_H - \xi \le \left[a_{HM}^M + (1-\alpha)a_{HM}^U\right]\omega_M \\
&\omega_H + \xi \ge \left[a_{HM}^M - (1-\alpha)a_{HM}^L\right]\omega_M \\
&\omega_I - \xi \le \left[a_{IF}^M + (1-\alpha)a_{IF}^U\right]\omega_F \\
&\omega_I + \xi \ge \left[a_{IF}^M - (1-\alpha)a_{IF}^L\right]\omega_F \\
&\omega_M - \xi \le \left[a_{MF}^M + (1-\alpha)a_{MF}^U\right]\omega_F \\
&\omega_M + \xi \ge \left[a_{MF}^M - (1-\alpha)a_{MF}^L\right]\omega_F \\
&\omega_H + \omega_F + \omega_I + \omega_M = 1 \\
&\omega_H, \omega_F, \omega_I, \omega_M \ge 0
\end{aligned}
\tag{4.13}
$$

After substituting the data of the elements in the BO and OW vectors, it could be obtained that:

$$
\begin{aligned}
&\min \xi \\
&s.t. \\
&\omega_H - \xi \le \left[3 + (1-\alpha)\frac{7}{2}\right]\omega_F \\
&\omega_H + \xi \ge \left[3 - (1-\alpha)\frac{5}{2}\right]\omega_F \\
&\omega_H - \xi \le \left[2 + (1-\alpha)\frac{5}{2}\right]\omega_I \\
&\omega_H + \xi \ge \left[2 - (1-\alpha)\frac{3}{2}\right]\omega_I \\
&\omega_H - \xi \le \left[1 + (1-\alpha)\frac{3}{2}\right]\omega_M \\
&\omega_H + \xi \ge \left[1 - (1-\alpha)\frac{2}{3}\right]\omega_M \\
&\omega_I - \xi \le \left[2 + (1-\alpha)\frac{5}{2}\right]\omega_F \\
&\omega_I + \xi \ge \left[2 - (1-\alpha)\frac{3}{2}\right]\omega_F \\
&\omega_M - \xi \le \left[3 + (1-\alpha)\frac{7}{2}\right]\omega_F \\
&\omega_M + \xi \ge \left[3 - (1-\alpha)\frac{5}{2}\right]\omega_F \\
&\omega_H + \omega_F + \omega_I + \omega_M = 1 \\
&\omega_H, \omega_F, \omega_I, \omega_M \ge 0
\end{aligned}
\tag{4.14}
$$

**Table 4.4** The independent weights of the four categories

| | $\omega_H$ | $\omega_F$ | $\omega_I$ | $\omega_M$ |
|---|---|---|---|---|
| $\alpha = 0.5$ | 0.3115 | 0.0984 | 0.1230 | 0.4672 |

The predefined possibility level $\alpha$ was set as 0.50 in this study, then, the independent weights of the four aspects can be obtained, and the results were presented in Table 4.4.

In a similar way, the inner dependency matrix can also be determined by FBWM. Taking the effects of the facilities aspect, institution aspect and management aspect on human aspect as an example, and the results were presented in Table 4.5.

In a similar way, all the elements in the inner dependency matrix can be determined, and the results were also presented in Table 4.6. The inner dependency matrix can then be determined, and the results were presented in Eq. (4.15).

$$\begin{vmatrix} & H & F & I & M \\ H & 1 & 0.3137 & 0.2027 & 0.4308 \\ F & 0.1387 & 1 & 0.1387 & 0.2462 \\ I & 0.2027 & 0.3922 & 1 & 0.3231 \\ M & 0.6587 & 0.2941 & 0.6587 & 1 \end{vmatrix} \tag{4.15}$$

According to Eq. (4.10), the inter-dependent weights of the four categories can be determined, and the results were presented in Eq. (4.16).

$$\begin{vmatrix} & H & F & I & M \\ H & 1 & 0.3137 & 0.2027 & 0.4308 \\ F & 0.1387 & 1 & 0.1387 & 0.2462 \\ I & 0.2027 & 0.3922 & 1 & 0.3231 \\ M & 0.6587 & 0.2941 & 0.6587 & 1 \end{vmatrix} \times \begin{vmatrix} 0.3115 \\ 0.0984 \\ 0.1230 \\ 0.4672 \end{vmatrix} = \begin{vmatrix} 0.5686 \\ 0.2737 \\ 0.3757 \\ 0.7823 \end{vmatrix} \tag{4.16}$$

Then, the normalized inter-dependent weights of the four categories can be determined according to Eq. (4.11), and the results were presented in Eqs. (4.17)–(4.20).

$$\frac{0.5686}{0.5686 + 0.2737 + 0.3757 + 0.7823} = 0.2843 \tag{4.17}$$

**Table 4.5** The effects of the facilities aspect, institution aspect and management aspect on human aspect

| | The most important: M | | The least important: F |
|---|---|---|---|
| Human aspect (H) | Facilities aspect (F) | Institution aspect (I) | Management aspect (M) |
| BO | (5/2, 3, 7/2) | (3/2, 2, 5/2) | (1, 1, 1) |
| OW | (1, 1, 1) | (3/2, 2, 5/2) | (5/2, 3, 7/2) |
| Weights | 0.1387 | 0.2027 | 0.6587 |

**Table 4.6** The effects of all the aspects on each aspect

| Facilities aspect (F) | The most important: I | | The least important: H |
|---|---|---|---|
| | Human aspect (H) | Institution aspect (I) | Management aspect (M) |
| BO | (3/2, 2, 5/2) | (1, 1, 1) | (2/3, 1, 3/2) |
| OW | (1, 1, 1) | (3/2, 2, 5/2) | (3/2, 2, 5/2) |
| Weights | 0.3137 | 0.3922 | 0.2941 |
| Institution aspect (I) | The most important: M | | The least important: F |
| | Human aspect (H) | Facilities aspect (F) | Management aspect (M) |
| BO | (3/2, 2, 5/2) | (5/2, 3, 7/2) | (1, 1, 1) |
| OW | (3/2, 2, 5/2) | (1, 1, 1) | (5/2, 3, 7/2) |
| Weights | 0.2027 | 0.1387 | 0.6587 |
| Management aspect (M) | The most important: H | | The least important: F |
| | Human aspect (H) | Facilities aspect (F) | Institution aspect (I) |
| BO | (1,1,1) | (5/2,3,7/2) | (2/3,1,3/2) |
| OW | (5/2,3,7/2) | (1,1,1) | (3/2,2,5/2) |
| Weights | 0.4308 | 0.2462 | 0.3231 |

$$\frac{0.2737}{0.5686 + 0.2737 + 0.3757 + 0.7823} = 0.1368 \quad (4.18)$$

$$\frac{0.3757}{0.5686 + 0.2737 + 0.3757 + 0.7823} = 0.1878 \quad (4.19)$$

$$\frac{0.7823}{0.5686 + 0.2737 + 0.3757 + 0.7823} = 0.3911 \quad (4.20)$$

After determining the normalized inter-dependent weights of the four categories, the independent weights of the criteria in each category can also be determined by the FBWM, and the results were presented in Table 4.7.

The final inter-dependent weights of the criteria can be determined by calculating the product of the independent weight of the criterion and the normalized inter-dependent weight of the corresponding category. Taking the final inter-dependent weights of 'Safety maintenance style and awareness ($H_1$)' as an example: the independent weight of $H_1$ × the normalized inter-dependent weight of human aspect (H) $= 0.2843 \times 0.5556 = 0.1580$. In a similar way, all the final inter-dependent weights of the criteria can be determined, and the results were presented in Table 4.8.

According to the results presented in Table 4.8, these criteria can be categorized into three groups: significantly important group, moderately important group, and weakly important group. Appropriate plan and schedule ($M_2$) and safety maintenance

**Table 4.7** The independent weights of the criteria in each category

| Human aspect (H) | The most important: $H_1$ | | The least important: $H_2$ |
|---|---|---|---|
| | Safety maintenance style and awareness ($H_1$) | Excellent collaborations and communication ($H_2$) | Excellent and mature operation skills ($H_3$) |
| BO | (1, 1, 1) | (2, 5/2, 3) | (1, 3/2, 2) |
| OW | (2, 5/2, 3) | (1, 1, 1) | (1, 3/2, 2) |
| Weights | 0.5556 | 0.2222 | 0.2222 |
| Facilities aspect (F) | The most important: $F_2$ | | The least important: $F_1$ |
| | Appropriate use of maintenance facilities ($F_1$) | Completion of factories and equipment ($F_2$) | Sufficient supply of aircraft materials ($F_3$) |
| BO | (5/2, 3, 7/2) | (1, 1, 1) | (1, 3/2, 2) |
| OW | (1, 1, 1) | (5/2, 3, 7/2) | (2, 5/2, 3) |
| Weights | 0.1307 | 0.6209 | 0.2484 |
| Institution aspect (I) | The most important: $I_1$ | | The least important: $I_3$ |
| | Regulations and standards ($I_1$) | Appropriate engineering and technological documents ($I_2$) | Working environment ($I_3$) |
| BO | (1, 1, 1) | (2/3, 1, 3/2) | (3/2, 2, 5/2) |
| OW | (3/2, 2, 5/2) | (1, 3/2, 2) | (1, 1, 1) |
| Weights | 0.5322 | 0.3041 | 0.1637 |
| Institution aspect (M) | The most important: $M_2$ | | The least important: $M_1$ |
| | Safety management system ($M_1$) | Appropriate plan and schedule ($M_2$) | Complete labor training system ($M_3$) |
| BO | (2, 5/2, 3) | (1, 1, 1) | (3/2, 2, 5/2) |
| OW | (1, 1, 1) | (2, 5/2, 3) | (1, 3/2, 2) |
| Weights | 0.1905 | 0.6190 | 0.1905 |

**Table 4.8** The final inter-dependent weights of the criteria

| $H_1$ | $H_2$ | $H_3$ | $F_1$ | $F_2$ | $F_3$ |
|---|---|---|---|---|---|
| 0.1580 | 0.0632 | 0.0632 | 0.0179 | 0.0849 | 0.0340 |
| $I_1$ | $I_2$ | $I_3$ | $M_1$ | $M_2$ | $M_3$ |
| 0.0999 | 0.0571 | 0.0307 | 0.0745 | 0.2421 | 0.0745 |

style and awareness ($H_1$) belong to the significantly important group. Excellent collaborations and communication ($H_2$), excellent and mature operation skills ($H_3$), completion of factories and equipment ($F_2$), regulations and standards ($I_1$), appropriate engineering and technological documents ($I_2$), safety management system ($M_1$), and complete labor training system ($M_3$) belong to the moderately important group. The other factors belong to the weakly important group.

According to the factors belonging to the significantly important group and moderately important group, the following strategic measures were proposed for assuring the aviation maintenance safety, and they are:

(1) Drafting appropriate plan and schedule ($M_1$);
(2) Education on safety maintenance awareness ($M_2$);
(3) Building teamwork mechanism for aviation maintenance ($M_3$);
(4) Training on aviation maintenance ($M_4$);
(5) Improving the infrastructure and facility conditions ($M_5$):
(6) Perfect the regulation and standard system ($M_6$);
(7) Improving the engineering and technological documents ($M_7$);
(8) Establishing complete safety management system ($M_8$).

The ISM method was employed to investigate the complex relationships among the strategic measures for improving the aviation maintenance safety, and the procedures were presented in Fig. 4.1.

The structural self-interaction matrix was firstly determined based on brainstorm by using 'V', 'A', 'X', and 'O' to represent the relationship of a strategic measure over another, and the results were presented in Table 4.9.

After this, the initial reachability matrix can be determined, and the results were presented in Table 4.10.

Subsequently, the final reachability matrix can be obtained according to Eq. (4.13), and the results were presented in Table 4.11.

According to Table 4.11, the driving power and dependence power of each strategic measure can be determined, and these eight strategic measures can be arranged

**Table 4.9** The structural self-interaction matrix for identifying the complex relationships among the eight strategic measures

| Measures | $M_8$ | $M_7$ | $M_6$ | $M_5$ | $M_4$ | $M_3$ | $M_2$ | $M_1$ |
|---|---|---|---|---|---|---|---|---|
| $M_1$ | V | O | O | V | O | O | O | – |
| $M_2$ | O | O | X | O | A | O | – | |
| $M_3$ | O | A | A | O | A | – | | |
| $M_4$ | O | V | O | O | – | | | |
| $M_5$ | A | A | O | – | | | | |
| $M_6$ | V | X | – | | | | | |
| $M_7$ | A | – | | | | | | |
| $M_8$ | – | | | | | | | |

**Table 4.10** The initial reachability matrix

| Measures | $M_1$ | $M_2$ | $M_3$ | $M_4$ | $M_5$ | $M_6$ | $M_7$ | $M_8$ |
|---|---|---|---|---|---|---|---|---|
| $M_1$ | 1 | 0 | 0 | 0 | 1 | 0 | 0 | 1 |
| $M_2$ | 0 | 1 | 0 | 0 | 0 | 1 | 0 | 0 |
| $M_3$ | 0 | 0 | 1 | 0 | 0 | 0 | 0 | 0 |
| $M_4$ | 0 | 1 | 1 | 1 | 0 | 0 | 1 | 0 |
| $M_5$ | 0 | 0 | 0 | 0 | 1 | 0 | 0 | 0 |
| $M_6$ | 0 | 1 | 1 | 0 | 0 | 1 | 1 | 1 |
| $M_7$ | 0 | 0 | 1 | 0 | 1 | 0 | 1 | 0 |
| $M_8$ | 0 | 0 | 0 | 0 | 1 | 1 | 1 | 1 |

**Table 4.11** The final reachability matrix

| Measures | $M_1$ | $M_2$ | $M_3$ | $M_4$ | $M_5$ | $M_6$ | $M_7$ | $M_8$ | Driving |
|---|---|---|---|---|---|---|---|---|---|
| $M_1$ | 1 | 1* | 1* | 0 | 1 | 1* | 1* | 1 | 7 |
| $M_2$ | 0 | 1 | 1* | 0 | 1 | 1 | 1* | 1* | 6 |
| $M_3$ | 0 | 0 | 1 | 0 | 0 | 0 | 0 | 0 | 1 |
| $M_4$ | 0 | 1 | 1 | 1 | 1* | 1* | 1 | 1* | 7 |
| $M_5$ | 0 | 0 | 0 | 0 | 1 | 0 | 0 | 0 | 1 |
| $M_6$ | 0 | 1 | 1 | 0 | 1* | 1 | 1 | 1 | 6 |
| $M_7$ | 0 | 0 | 1 | 0 | 1 | 0 | 1 | 0 | 3 |
| $M_8$ | 0 | 1 | 1 | 0 | 1 | 1 | 1 | 1 | 6 |
| Dependence | 1 | 5 | 7 | 1 | 7 | 5 | 6 | 5 | |

*Incorporating transitivity

into four quadrants, namely quadrant I (autonomous quadrant), quadrant II (dependent quadrant), quadrant III (linkage quadrant), and quadrant IV(independent quadrant), as presented in Fig. 4.2. It is apparent that there are not any measures belonging to quadrant I (autonomous quadrant), and it means that there is no measure which has the weak driving power and dependence power, and all the measures are highly connected. Building teamwork mechanism for aviation maintenance ($M_3$), improving the infrastructure and facility conditions ($M_5$), and improving the engineering and technological documents ($M_7$) belong to quadrant II (dependent quadrant), and it means that these strategic measures for improving aviation maintenance safety have relatively weak driving power but relatively strong dependence power, and these three strategic measures were highly affected by other measures but have few influences on the other strategic measures. Education on safety maintenance awareness ($M_2$), perfect the regulation and standard system ($M_6$), and establishing complete safety management system ($M_8$) belong to quadrant III (linkage quadrant), it means that these measures have both relatively strong driving power and strong dependence power, but they are unstable. In other words, an influence on these measures may lead

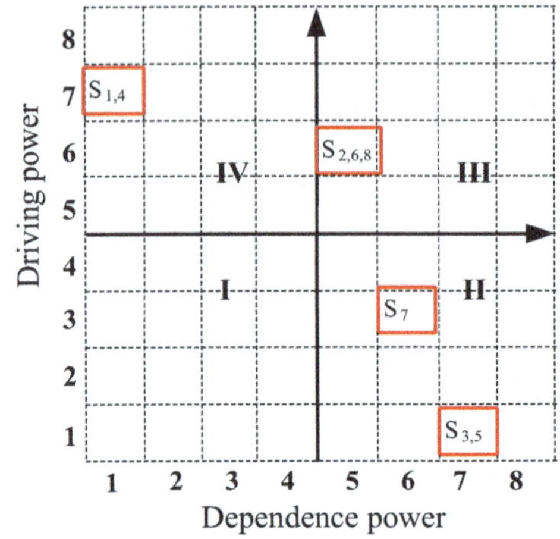


**Fig. 4.2** Driving power and dependence power diagram

the quadrants which they belong to change. Drafting appropriate plan and schedule ($M_1$) and training on aviation maintenance ($M_4$) belong to quadrant IV (independent quadrant), and these strategic measures have weak dependence power but strong driving power, and they are the critical measures that have the most significant effects on improving aviation maintenance safety.

Then, the reachability set, the antecedent set, and the intersection set with respect to each strategic measure can be determined, as presented in Table 4.12, and it is obvious that the reachability set and the intersection set with respect to building teamwork mechanism for aviation maintenance ($M_3$) and improving the infrastructure and facility conditions ($M_5$) are the same, thus, the strategic measures' 'building teamwork mechanism for aviation maintenance' and improving the infrastructure

**Table 4.12** Level partition of the factors-Iteration 1

| | Reachability set | Antecedent set | Intersection set | Level |
|---|---|---|---|---|
| $M_1$ | $M_1$, $M_2$, $M_3$, $M_5$, $M_6$, $M_7$, $M_8$ | $M_1$ | $M_1$ | |
| $M_2$ | $M_2$, $M_3$, $M_5$, $M_6$, $M_7$, $M_8$ | $M_1$, $M_2$, $M_4$, $M_6$, $M_8$ | $M_2$, $M_6$, $M_8$ | |
| $M_3$ | $M_3$ | $M_1$, $M_2$, $M_3$, $M_4$, $M_6$, $M_7$, $M_8$ | $M_3$ | I |
| $M_4$ | $M_2$, $M_3$, $M_4$, $M_5$, $M_6$, $M_7$, $M_8$ | $M_4$ | $M_4$ | |
| $M_5$ | $M_5$ | $M_1$, $M_2$, $M_4$, $M_5$, $M_6$, $M_7$, $M_8$ | $M_5$ | I |
| $M_6$ | $M_2$, $M_3$, $M_5$, $M_6$, $M_7$, $M_8$ | $M_1$, $M_2$, $M_4$, $M_6$, $M_8$ | $M_2$, $M_6$, $M_8$ | |
| $M_7$ | $M_3$, $M_5$, $M_7$ | $M_1$, $M_2$, $M_4$, $M_6$, $M_7$, $M_8$ | $M_7$ | |
| $M_8$ | $M_2$, $M_3$, $M_5$, $M_6$, $M_7$, $M_8$ | $M_1$, $M_2$, $M_4$, $M_6$, $M_8$ | $M_2$, $M_6$, $M_8$ | |

and facility conditions' should be positioned in the top level (level-I). After determining the measure in the top level, $M_3$ and $M_5$ were discarded from the measure list. In a similar way, the strategic measures belonging to level-II can also be determined. Similarly, the measures in the other levels can also be determined, and the results were presented in Tables 4.13, 4.14 and 4.15. The results can be described as follows: building teamwork mechanism for aviation maintenance ($M_3$) and improving the infrastructure and facility conditions ($M_5$) belong to level I; improving the engineering and technological documents ($M_7$) belongs to level II; education on safety maintenance awareness ($M_2$), perfect the regulation and standard system ($M_6$), and establishing complete safety management system ($M_8$) belong to level III; and drafting appropriate plan and schedule ($M_1$) and training on aviation maintenance ($M_4$) belong to level IV.

The hierarchical model for describing the complex relationships between two pair of the strategic measures could be obtained, and the results were presented in Fig. 4.3.

According to Fig. 4.3, drafting appropriate plan and schedule and training on aviation maintenance are the foundation for improving aviation maintenance safety.

**Table 4.13** Level partition of the factors-Iteration 2

| | Reachability set | Antecedent set | Intersection set | Level |
|---|---|---|---|---|
| $M_1$ | $M_1$, $M_2$, $M_6$, $M_7$, $M_8$ | $M_1$ | $M_1$ | |
| $M_2$ | $M_2$, $M_6$, $M_7$, $M_8$ | $M_1$, $M_2$, $M_4$, $M_6$, $M_8$ | $M_2$, $M_6$, $M_8$ | |
| $M_4$ | $M_2$, $M_4$, $M_6$, $M_7$, $M_8$ | $M_4$ | $M_4$ | |
| $M_6$ | $M_2$, $M_6$, $M_7$, $M_8$ | $M_1$, $M_2$, $M_4$, $M_6$, $M_8$ | $M_2$, $M_6$, $M_8$ | |
| $M_7$ | $M_7$ | $M_1$, $M_2$, $M_4$, $M_6$, $M_7$, $M_8$ | $M_7$ | II |
| $M_8$ | $M_2$, $M_6$, $M_7$, $M_8$ | $M_1$, $M_2$, $M_4$, $M_6$, $M_8$ | $M_2$, $M_6$, $M_8$ | |

**Table 4.14** Level partition of the factors-Iteration 3

| | Reachability set | Antecedent set | Intersection set | Level |
|---|---|---|---|---|
| $M_1$ | $M_1$, $M_2$, $M_6$, $M_8$ | $M_1$ | $M_1$ | |
| $M_2$ | $M_2$, $M_6$, $M_8$ | $M_1$, $M_2$, $M_4$, $M_6$, $M_8$ | $M_2$, $M_6$, $M_8$ | III |
| $M_4$ | $M_2$, $M_4$, $M_6$, $M_8$ | $M_4$ | $M_4$ | |
| $M_6$ | $M_2$, $M_6$, $M_8$ | $M_1$, $M_2$, $M_4$, $M_6$, $M_8$ | $M_2$, $M_6$, $M_8$ | III |
| $M_8$ | $M_2$, $M_6$, $M_8$ | $M_1$, $M_2$, $M_4$, $M_6$, $M_8$ | $M_2$, $M_6$, $M_8$ | III |

**Table 4.15** Level partition of the factors-Iteration 4

| | Reachability set | Antecedent set | Intersection set | Level |
|---|---|---|---|---|
| $M_1$ | $M_1$ | $M_1$ | $M_1$ | IV |
| $M_4$ | $M_4$ | $M_4$ | $M_4$ | IV |

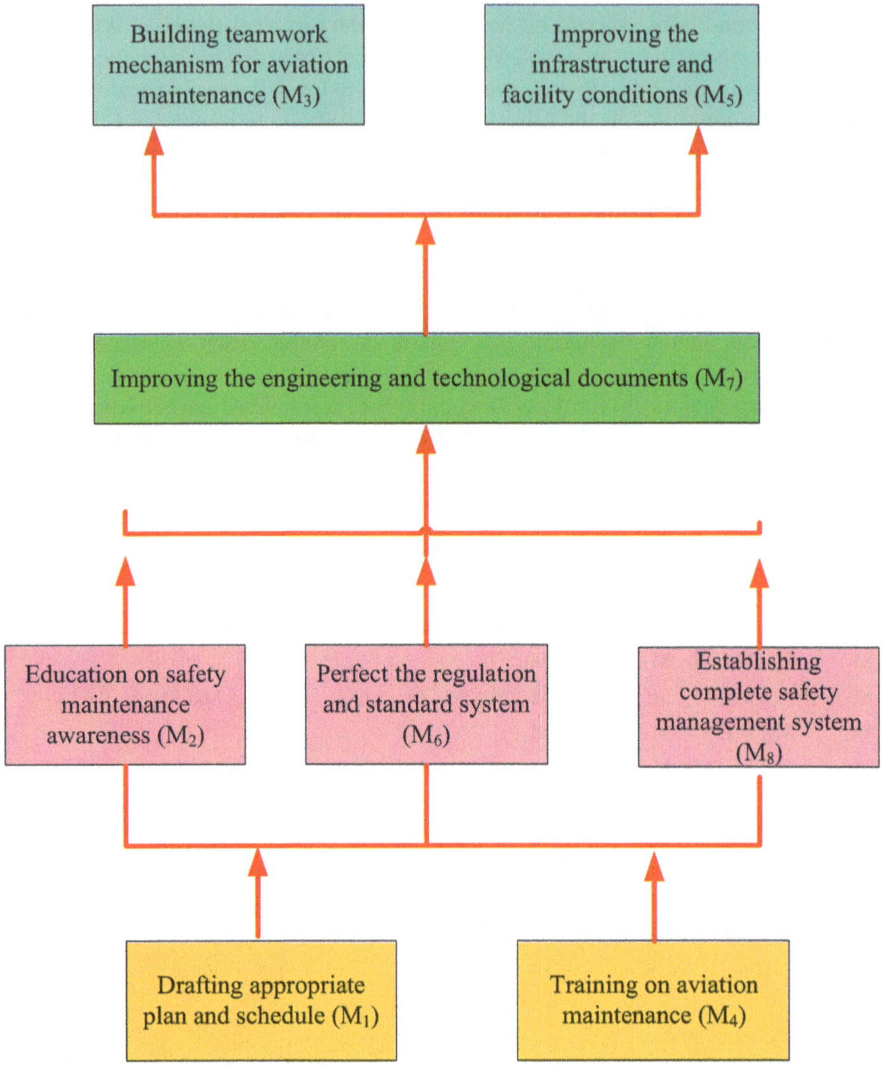


**Fig. 4.3** ISM based hierarchical model for strategic measures for improving the aviation maintenance safety

Education on safety maintenance awareness, perfect the regulation and standard system, and establishing complete safety management system are the three main pillars of safe aviation maintenance.

## 4.4 Conclusions

This study aims at employing the multi-criteria decision analysis methods for identifying the critical enablers for aviation maintenance safety. The enablers of aviation maintenance safety were firstly summarized based on literature reviews and focus group meetings; subsequently, the fuzzy best-worst network method was developed to determine the relative importance (weights) of the enablers and prioritize these enablers; then, the strategic measures for improving and ensuring aviation maintenance safety were proposed based on the critical enablers; finally, the Interpretative Structural Modeling was employed to investigate the complex relationships among these strategic measures. The results reveal that the decision-makers/stakeholders should pay more attentions on the following five strategic:

(1) drafting appropriate plan and schedule;
(2) training on aviation maintenance;
(3) education on safety maintenance awareness;
(4) perfect the regulation and standard system;
(5) and establishing complete safety management system.

## References

F. Abadi, I. Sahebi, A. Arab, A. Alavi, H. Karachi, Application of best-worst method in evaluation of medical tourism development strategy. Decis. Sci. Lett. **7**(1), 77–86 (2018)

M.A. Agi, R. Nishant, Understanding influential factors on implementing green supply chain management practices: An interpretive structural modelling analysis. J. Environ. Manage. **188**, 351–363 (2017)

M. Dağdeviren, İ. Yüksel, M. Kurt, A fuzzy analytic network process (ANP) model to identify faulty behavior risk (FBR) in work system. Saf. Sci. **46**(5), 771–783 (2008)

M.R. Endsley, M.M. Robertson, Situation awareness in aircraft maintenance teams. Int. J. Ind. Ergon. **26**(2), 301–325 (2000)

Gao and Wang, Research on influence factors of maintenance safety in general aviation based on G1-DEMATEL method. J. Saf. Sci. Technol. **12**(2), 164–169 (2016). (in Chinese)

J. Gao, M. Duan, L. Zhao, Q. Che, Aviation equipment maintenance and support safety risk assessment based on improved ANP. Aviat. Maint Eng **5**, 58–60 (2010). (in Chinese)

S. Guo, H. Zhao, Fuzzy best-worst multi-criteria decision-making method and its applications. Knowl.-Based Syst. **121**, 23–31 (2017)

H. Gupta, M.K. Barua, Supplier selection among SMEs on the basis of their green innovation ability using BWM and fuzzy TOPSIS. J. Clean. Prod. **152**, 242–258 (2017)

A. Hafezalkotob, A. Hafezalkotob, A novel approach for combination of individual and group decisions based on fuzzy best-worst method. Appl. Soft Comput. (2017)

G. Kannan, S. Pokharel, P.S. Kumar, A hybrid approach using ISM and fuzzy TOPSIS for the selection of reverse logistics provider. Resour. Conserv. Recycl. **54**(1), 28–36 (2009)

C.Y. Kim, B.H. Song, A study on safety culture in aviation maintenance organization. Adv. Sci. Technol. Lett. **120**, 485–490 (2015)

R.M. Knotts, Civil aircraft maintenance and support fault diagnosis from a business perspective. J. Qual. Maint. Eng. **5**(4), 335–348 (1999)

J. Rezaei, Best-worst multi-criteria decision-making method. Omega **53**, 49–57 (2015)

J. Ren, S. Tan, M.E. Goodsite, B.K. Sovacool, L. Dong, Sustainability, shale gas, and energy transition in China: assessing barriers and prioritizing strategic measures. Energy **84**, 551–562 (2015)

J. Ren, H. Liang, F.T. Chan, Urban sewage sludge, sustainability, and transition for Eco-City: Multi-criteria sustainability assessment of technologies based on best-worst method. Technol. Forecast. Soc. Change **116**, 29–39 (2017)

J.N. Warfield, Toward interpretation of complex structural modeling. IEEE Trans. Syst. Man Cybern. **4**(5), 405–417 (1974a)

J.N. Warfield, Developing interconnection matrices in structural modeling. IEEE Trans. Syst. Man Cybern. **4**(1), 81–87 (1974b)

Y. Xia, D. Guo, H. Zhang, Research on assessment parameters of aviation service safety culture. Aviat. Maint Eng. **5**, 77–80 (2010). (in Chinese)

M. Zamani, A. Rabbani, A. Yazdani-Chamzini, Z. Turskis, An integrated model for extending brand based on fuzzy ARAS and ANP methods. J. Bus. Econ. Manag. **15**(3), 403–423 (2014)

Y. Zhang, S. Wu, X. Liu, B. He, J. Xiao, Dynamic evaluation of aviation maintenance safety based on set pair analysis and Marko chain. China Saf. Sci. J. **26**(1), 122–128 (2016). (in Chinese)

# Chapter 5
# Multi-stakeholder Multi-criteria Decision-Making Framework for Sustainability Prioritization: Investigation of the Processes for Sludge-to-Wealth

**Abstract** The treatment of sewage sludge has been a great challenge of many counties. This study aims at developing a generic multi-stakeholder multi-criteria decision making framework for sustainability assessment and prioritization of the pathways for sludge recycling. Multi-stakeholder intuitionistic fuzzy analytic hierarchy process (AHP) method which allows multiple groups of stakeholders to participate in the determination of the weights of the evaluation criteria was developed. Multi-stakeholder intuitionistic fuzzy grey relational analysis (GRA) method was proposed to rank the alternative processes for converting sludge to energy or resources. Four pathways for sludge treatment including sludge for landscaping, sludge for landfilling, sludge for incineration, and sludge for cement production were studied, and this method enables the decision-makers to know the integrated priorities of the processes for sludge treatment. Sludge for landscaping has been recognized as the most sustainable pathway for sludge treatment, follows by sludge for landfilling, sludge for cement production ($A_4$), and sludge for incineration. The results were validated by the sum weighted method, and sensitivity analysis was also carried to investigate the effects of the weights on the final sustainability ranking.

## 5.1 Introduction

With the economic prosperity, rapid industrialization, and the increase of urban population, sewage sludge as the residual of wastewater treatment has become a great challenge recently. The treatment of sewage sludge has become an urgent task of many counties (Ren et al. 2017). There are usually various ways for the treatment of sewage sludge, including landfill, incineration, pyrolysis, gasification, and using as material source for cement production, etc. (Chen et al. 2012). Among these methods, sludge recycling utilization attracts more and more attentions, because it cannot only mitigate the negative environmental impacts, but also lower the risks of resource scarcity and energy security.

As mentioned above, sludge recycling utilization option has been recognized as a promising way for handling sewage sludge, thus various studies have been carried to investigate the priorities and superiorities of these scenarios. Fersi et al. (2015)

 
J. Ren, *Advanced Operations Management for Complex Systems Analysis*,
SpringerBriefs in Applied Sciences and Technology,
https://doi.org/10.1007/978-3-030-45418-0_5

investigated the economic performances of renewable heat and electricity production by sewage sludge digestion. Akwo (2008) has investigated the life cycle environmental impacts of different sewage sludge treatment options, and the results reveal that different technologies for urban sludge treatment perform different. Lederer and Rechberger (2010) compared different traditional pathways and an innovative thermo-chemical process to treat and dispose of sewage sludge with regard to environmental impact, resource recovery, and materials dissipation. Eriksson et al. (2016) studied the environmental impacts and economic benefits of biogas production from food waste and sewage sludge in life cycle perspective. There are also some other studies focusing on environmental assessment and comparisons of different sludge disposal options, i.e. anaerobic digestion versus thermal processes (Hospido et al. 2005), supercritical water oxidation versus other sewage sludge handling options (i.e. agricultural use and co-incineration with municipal solid waste) (Svanström et al. 2005). All these studies find that different options for sewage sludge treatment have different performances in different aspects. Accordingly, it is a challenging task for the decision-makers to select the best scenario for sewage sludge treatment among multiple alternatives.

Strategic decision making on selecting the most suitable pathway for converting sludge into energy or resources among multiple alternatives which is similar to the selection waste-to-energy technology of is usually difficult and complex, and the operations research tools such as multi-attribute decision analysis (MADA) methods are usually adopted for helping the decision-makers to have a comprehensive understanding of these alternatives (Yap and Nixon 2015). MADA methods have been widely used for energy engineering, especially for the selection of the best energy system among multiple alternatives, i.e. multi-criteria assessment of hybrid energy system (Diemuodeke et al. 2016), multi-criteria evaluation of wave energy project (Flocard et al. 2016), multi attribute decision model for compressor selection (Taylan et al. 2016), ranking the renewables for electricity generation (Haddad et al. 2017), and location selection for wind farms (Gigović et al. 2017). MADA is a powerful tool for ranking the alternatives according to their performances with respect to some evaluation criteria; however, the traditional MADA methods have to know the exact data of the alternatives with respect to the evaluation criteria. While some data of the sludge recycling pathways with respect to the evaluation criteria cannot be obtained due to the lack of information/knowledge, thus, a MADA method based on the intuitionistic fuzzy set theory was developed in this study for sustainability assessment and prioritization of the sludge treatment processes. In order to incorporate the opinions of all the stakeholders related to sludge treatment, the MCDA method has been extended to multi-stakeholder decision making condition, and the proposed multi-stakeholder MADA method was employed to assess the sustainability of sludge treatment processes in this study.

Besides this section, the evaluation criteria for sustainability assessment of the processes for handling sewage sludge was presented in Sects. 5.2 and 5.3 developed the multi-stakeholder multi-attribute decision analysis for prioritizing the processes for handling sewage sludge; a case has been studied for illustrating the developed model in Sect. 5.4; the results were discussed through validation by another MADA method and sensitivity analysis in Sect. 5.5; finally, this study has been conclude in Sect. 5.6.

## 5.2 Evaluation Criteria

Most of the studies about the evaluation of the alternative pathways for converting sludge to resources merely focused on technological and economic performances; however, Pohoo-Aikins et al. (2010) pointed out that it is necessary to have a consistent basis for comparing these alternatives based on key criteria. In order to have a comprehensive evaluation of the alternative pathways for sludge-to-resources, the concept of sustainability has been incorporated in the evaluation. Sustainability assessment which aims at assessing the processes/products comprehensively with the considerations of economic prosperity, environmental cleanness, social acceptance, and technological priority has been used for the selection of the most sustainable pathways for converting sludge to resources.

A total of nine criteria in four categories, including economic performance, environmental impact, social effect, and technological were employed to assess the sustainability of these pathways (see Table 5.1). Capital cost, operation cost and benefit were used to measure economic performance; environmental category consists of the emissions of $NO_x$ and $SO_2$; social category includes added jobs and compliance with the policy; there are two criteria in technological category, and they are technology maturity and management requirement.

**Table 5.1** The criteria for sustainability assessment of the sludge treatment pathways (Meng 2015)

| Category | Criteria | Abbreviation |
|---|---|---|
| | Capital cost | $EC_1$ |
| Economic (EC) | Operation cost | $EC_2$ |
| | Benefit | $EC_3$ |
| | $NO_x$ emission | $EN_1$ |
| Environmental (EN) | $SO_2$ emission | $EN_2$ |
| Social (S) | Added jobs | $S_1$ |
| | Compliance with the policy | $S_2$ |
| Technological (T) | Technology maturity | $T_1$ |
| | Management requirement | $T_2$ |

## 5.3 Methods

Section 5.3.1 illustrates the intuitionistic fuzzy set theory and the preliminaries of fuzzy set theory; multi-stakeholder intuitionistic fuzzy AHP method was developed in Sect. 5.3.2, and the multi-stakeholder intuitionistic fuzzy grey relational analysis was demonstrated in Sect. 5.3.3.

### *5.3.1 Preliminaries*

The fuzzy set theory developed by Zadeh (2015) was defined as follows:

**Definition 1** Fuzzy sets (FS)

Imaging that X refers to the universe of discourse, which is composed of x, and the membership function $\mu_{\tilde{\alpha}}(x)$ can characterize a fuzz set $\tilde{\alpha}$, which the x's degree of belonging to α. The degree of the membership of x in $\tilde{\alpha}$ can be indicated by the value of $\mu_{\tilde{\alpha}}(x)$.

$$\alpha = \{(x, \mu_{\alpha}(x))|x \in X\} \tag{5.1}$$

**Definition 2** Intuitionistic fuzzy set (IFS)

Assuming that X is an object collection of x and $\alpha \in$ X refers to the fixed set, an intuitionistic fuzzy set $\alpha$ on X can be defined as (Yuan et al. 2014):

$$\alpha = \{(x, \mu_{\alpha}(x), \upsilon_{\alpha}(x))|x \in X\} \tag{5.2}$$

in which, $\mu_{\alpha}(x) : X \rightarrow [0, 1]$, $x \in X \rightarrow \mu_{\alpha}(x) \in [0, 1]$ stands for the level of the membership of $x \in X$ to the set $\alpha$, and $\upsilon_{\alpha}(x)$:$X \rightarrow [0, 1]$, $x \in X \rightarrow \upsilon_{\alpha}(x) \in [0, 1]$ refers to the level of the non-membership of the element $x \in X$ to the set $\alpha$.

$\mu_{\alpha}(x)$ and $\upsilon_{\alpha}(x)$ usually meet the criteria of $0 \le \mu_{\alpha}(x) + \upsilon_{\alpha}(x) \le 1$ for all $x \in X$. Apart from the level of both the membership and non-membership, x's hesitancy degree or the indeterminacy to the set $\alpha$, which differentiates the numbers $\mu_{\alpha}(x)$ and $\upsilon_{\alpha}(x)$ standing for the level of the membership and the level of the non-membership of the element $x \in X$ to the set $\alpha$, can gauge the indeterminacy level of $x \in X$ to the set $\alpha$ is defined as:

$$\pi_{\alpha}(x) = 1 - \mu_{\alpha}(x) - \upsilon_{\alpha}(x), \ x \in X \tag{5.3}$$

Correspondingly, $\alpha = (\mu_{\alpha}, \upsilon_{\alpha}, \pi_{\alpha})$ represents the intuitionistic fuzzy number $\alpha$ by involving the level of indeterminacy, non-membership, and level of membership.

**Definition 3** Arithmetic operations (Yu and Xu 2013; Chen 2014)

Imagine $\alpha = (\mu_\alpha, \upsilon_\alpha, \pi_\alpha)$ and $\beta = (\mu_\beta, \upsilon_\beta, \pi_\beta)$ are two distinctive fuzzy numbers. The two intuitionistic fuzzy numbers' arithmetic operations, which are demonstrated in below:

**Addition**

$$\begin{aligned} \alpha \oplus \beta &= (\mu_\alpha, \upsilon_\alpha, \pi_\alpha) \oplus (\mu_\beta, \upsilon_\beta, \pi_\beta) \\ &= (\mu_\alpha + \mu_\beta - \mu_\alpha\mu_\beta, \upsilon_\alpha\upsilon_\beta, 1 + \mu_\alpha\mu_\beta - \mu_\alpha - \mu_\beta - \upsilon_\alpha\upsilon_\beta) \end{aligned} \tag{5.4}$$

For example, $\alpha = (0.7, 0.2, 0.1)$ and $\beta = (0.3, 0.5, 0.2)$, Eq. (5.5) can determine the sum of two intuitionistic fuzzy numbers.

$$\begin{aligned} \alpha + \beta &= (0.7, 0.2, 0.1) + (0.3, 0.5, 0.2) \\ &= (0.7 + 0.3 - 0.7 \times 0.3, 0.2 \times 0.5, 1 + 0.7 \times 0.3 - 0.7 - 0.3 - 0.2 \times 0.5) \\ &= (0.79, 0.10, 0.11) \end{aligned} \tag{5.5}$$

$$\begin{aligned} \mathop{\oplus}_{j=1}^{n} \alpha_j &= \mathop{\oplus}_{j=1}^{n} (\mu_{\alpha_j}, \upsilon_{\alpha_j}, \pi_{\alpha_j}) \\ &= \left(1 - \prod_{j=1}^{n}(1 - \mu_{\alpha_j}), \prod_{j=1}^{n} \upsilon_{\alpha_j}, \prod_{j=1}^{n}(1 - \mu_{\alpha_j}) - \prod_{j=1}^{n} \upsilon_{\alpha_j}\right) \end{aligned} \tag{5.6}$$

For example, $\alpha_1 = (0.6, 0.3, 0.1)$, $\alpha_2 = (0.7, 0.2, 0.1)$, and $\alpha_3 = (0.8, 0.1, 0.1)$, then, it can be obtained that:

$$\begin{aligned} \mathop{\oplus}_{j=1}^{3} \alpha_j &= \alpha_1 \oplus \alpha_2 \oplus \alpha_3 \\ &= (1 - (1 - 0.6) \times (1 - 0.7) \times (1 - 0.8), 0.3 \times 0.1 \times 0.1 \\ &\quad (1 - 0.6) \times (1 - 0.7) \times (1 - 0.8) - 0.3 \times 0.1 \times 0.1) \\ &= (0.976, 0.003, 0.021) \end{aligned} \tag{5.7}$$

**Multiplication**

$$\begin{aligned} \alpha \otimes \beta &= (\mu_\alpha, \upsilon_\alpha, \pi_\alpha) \otimes (\mu_\beta, \upsilon_\beta, \pi_\beta) \\ &= (\mu_\alpha\mu_\beta, \upsilon_\alpha + \upsilon_\beta - \upsilon_\alpha\upsilon_\beta, 1 + \upsilon_\alpha\upsilon_\beta - \mu_\alpha\mu_\beta - \upsilon_\alpha - \upsilon_\beta) \end{aligned} \tag{5.8}$$

For example, $\alpha = (0.7, 0.2, 0.1)$ and $\beta = (0.3, 0.5, 0.2)$, the two intuitionistic fuzzy numbers' the product can be determined by Eq. (5.9).

$$\begin{aligned} \alpha \otimes \beta &= (0.7, 0.2, 0.1) \otimes (0.3, 0.5, 0.2) \\ &= (0.7 \times 0.3, 0.2 + 0.5 - 0.2 \times 0.5, 1 + 0.2 \times 0.5 - 0.7 \times 0.3 - 0.2 - 0.5) \\ &= (0.21, 0.60, 0.19) \end{aligned} \tag{5.9}$$

$$\bigotimes_{j=1}^{n} \alpha_j = \bigotimes_{j=1}^{n} \left(\mu_{\alpha_j}, \upsilon_{\alpha_j}, \pi_{\alpha_j}\right) = \left(\prod_{j=1}^{n} \mu_{\alpha_j}, \prod_{j=1}^{n}\left(1 - \upsilon_{\alpha_j}\right), 1 - \prod_{j=1}^{n} \mu_{\alpha_j} - \prod_{j=1}^{n}\left(1 - \upsilon_{\alpha_j}\right)\right) \tag{5.10}$$

For example, $\alpha_1 = (0.6, 0.3, 0.1)$, $\alpha_2 = (0.7, 0.2, 0.1)$, and $\alpha_3 = (0.8, 0.1, 0.1)$, then, it can be obtained that:

$$\begin{aligned} \bigotimes_{j=1}^{3} \alpha_j &= \bigotimes_{j=1}^{3} \left(\mu_{\alpha_j}, \upsilon_{\alpha_j}, \pi_{\alpha_j}\right) \\ &= (0.6 \times 0.7 \times 0.8, (1 - 0.3) \times (1 - 0.1) \times (1 - 0.1), \\ &\quad 1 - 0.6 \times 0.7 \times 0.8 - (1 - 0.3) \times (1 - 0.1) \times (1 - 0.1)) \\ &= (0.336, 0.567, 0.097) \end{aligned} \tag{5.11}$$

**Definition 4** Ranking the intuitionistic fuzzy numbers (Yue 2016)

Equations (5.12), (5.13) can be used for the ranking for the intuitionistic fuzzy numbers $\alpha = (\mu_\alpha, \upsilon_\alpha, \pi_\alpha)$ and $\beta = \left(\mu_\beta, \upsilon_\beta, \pi_\beta\right)$.

$$\rho(\alpha) = \frac{(1 + \pi_\alpha)(1 - \mu_\alpha)}{2} \tag{5.12}$$

$$\rho(\beta) = \frac{\left(1 + \pi_\beta\right)\left(1 - \mu_\beta\right)}{2} \tag{5.13}$$

(i) $\alpha$ is greater than $\beta$, denotes by $\alpha = (\mu_\alpha, \upsilon_\alpha, \pi_\alpha) > \beta = \left(\mu_\beta, \upsilon_\beta, \pi_\beta\right)$ when $\rho(\alpha) < \rho(\beta)$;

(ii) $\alpha$ is indifferent to $\beta$, denotes by $\alpha = (\mu_\alpha, \upsilon_\alpha, \pi_\alpha) = \beta = \left(\mu_\beta, \upsilon_\beta, \pi_\beta\right)$ when $\rho(\alpha) = \rho(\beta)$;

(iii) $\alpha$ is smaller than $\beta$, denotes by $\alpha = (\mu_\alpha, \upsilon_\alpha, \pi_\alpha) < \beta = \left(\mu_\beta, \upsilon_\beta, \pi_\beta\right)$ when $\rho(\alpha) > \rho(\beta)$;

For example, $\alpha = (0.7, 0.2, 0.1)$ and $\beta = (0.3, 0.5, 0.2)$, it can be obtained that:

$$\rho(\alpha) = \frac{(1 + 0.1)(1 - 0.7)}{2} = 0.0825 \tag{5.14}$$

$$\rho(\beta) = \frac{(1 + 0.2)(1 - 0.3)}{2} = 0.42 \tag{5.15}$$

It is apparent that $\rho(\alpha) = 0.0825 < \rho(\beta) = 0.42$, then, it could be deduced that $\alpha = (0.7, 0.2, 0.1) > \beta = (0.3, 0.5, 0.2)$.

### 5.3.2 Multi-stakeholder Intuitionistic Fuzzy AHP Method

The weights of the criteria play an important role in multi-criteria decision making, because it cannot only reflect the preferences of the decision-makers, but also denote the relative importance of the criteria. There are various weights methods, i.e. subjective weighting method, objective weighting methods, and the combined methods. Analytic Hierarchy Process as one of the subjective weighting methods for determining the weights is the most widely used (Saaty 1980). In order to overcome the drawbacks of AHP (i.e. the lack of the ability for solve the ambiguity and vagueness existing in the opinions of the decision-makers and the difficult for establishing a consistent comparison matrix). The multi-stakeholder intuitionistic fuzzy AHP method was employed in this study. It consists of six steps based on the work of Wei (2011):

**Step 1**: Determine the panel of stakeholders and the hierarchical structure. This objective of this step is to determine all the related stakeholders in the decision-making process to achieve group decision-making and establish the hierarchical structure of the decision-making problem which usually consists of objective, the dimensions/aspect in the first hierarchy and criteria/indicators in each dimension/aspect of the first hierarchy.

**Step 2**: Determine the comparison judgement matrix by each of the decision-makers. Assuming that there are n criteria $C_1, C_2, \cdots, C_n$ to be studied, each of the decision-makers will be asked to assess the relative priority/importance of the one criterion over another. Each of the decision-makers are firstly asked to use linguistic terms derived from AHP and intuitionistic fuzzy AHP (see Table 5.2) to establish comparison matrix (Abdullah and Najib 2014).

**Table 5.2** Intuitionistic fuzzy numbers, crisp numbers and their corresponding linguistic terms used in intuitionistic fuzzy AHP and AHP (Abdullah and Najib 2014)

| Preference | Abbreviation | AHP scales | Intuitionistic fuzzy numbers (IFNs) | Reciprocals of IFN |
|---|---|---|---|---|
| Equally important | E | 1 | (0.02, 0.18, 0.80) | (0.02, 0.18, 0.80) |
| Intermediate value | EM | 2 | (0.06, 0.23, 0.70) | (0.23, 0.06, 0.70) |
| Moderately more important | MM | 3 | (0.13, 0.27, 0.60) | (0.27, 0.13, 0.60) |
| Intermediate value | MS | 4 | (0.22, 0.28, 0.50) | (0.28, 0.22, 0.50) |
| Strongly more important | SM | 5 | (0.33, 0.27, 0.40) | (0.27, 0.33, 0.40) |
| Intermediate value | SI | 6 | (0.47, 0.23, 0.30) | (0.23, 0.47, 0.30) |
| Very strong more important | VS | 7 | (0.62, 0.18, 0.20) | (0.18, 0.62, 0.20) |
| Intermediate value | VE | 8 | (0.80, 0.10, 0.10) | (0.10, 0.80, 0.30) |
| Extremely more important | EM | 9 | (1.0, 0, 0) | (1, .0, 0) |

**Table 5.3** Linguistic variables for rating the role of the decision-makers (Zhang and Liu 2011)

| Linguistic terms | Abbreviation | IFN |
|---|---|---|
| Very important (VI) | VI | (0.90, 0.05, 0.05) |
| Important (I) | I | (0.75, 0.20, 0.05) |
| Medium (M) | M | (0.50, 0.40, 0.10) |
| Unimportant (U) | U | (0.25, 0.60, 0.15) |
| Very unimportant (VU) | VU | (0.10, 0.80, 0.10) |

Assuming that there are a total of K experts, and the comparison matrix determined by the k-th decision-maker ($k = 1, 2, \cdots, K$) was presented in Eq. 5.16.

$$\begin{array}{cccccc} & C_1 & C_2 & \cdots & C_n \\ C_1 & a_{11}^k & a_{12}^k & \cdots & a_{1n}^k \\ C_2 & a_{21}^k & a_{22}^k & \cdots & a_{2n}^k \\ \vdots & \vdots & \vdots & \ddots & \vdots \\ C_n & a_{n1}^k & a_{n2}^k & \cdots & a_{nn}^k \end{array} \tag{5.16}$$

where $a_{ij}^k = \left(\mu_{a_{ij}^k}, \upsilon_{a_{ij}^k}, \pi_{a_{ij}^k}\right)(j = 1, 2, \ldots, n)$ is the relative priority/importance of the i-th criterion $C_i$ to that of the j-th criterion $C_j$.

Accordingly, a total of K different comparison matrices can be obtained. It is apparent that when $i = j$, then $a_{ij} = (0.02, 0.18, 0.80)$.

**Step 3**: Determine the aggregated intuitionistic fuzzy comparison judgement matrix by aggregating the K different comparison matrices into an aggregated one. There are K decision-makers in the decision-making; however, they usually paly different roles in the decision-making, thus, they should have different weights in the decision-making.

The linguistic terms presented in Table 5.3 were used to depict the relative importance of the role of the decision-makings, and these linguistic can also be transformed into the corresponding intuitionistic fuzzy numbers.

Let $D_k = (\mu_k^D, \upsilon_k^D, \pi_k^D)$ be the $k$-th role importance of decision-maker by using the IFN presented in Table 5.1, and the crisp weight of the k-th decision-maker could be determined by Eq. 5.17 (Boran et al. 2009).

$$\lambda_k = \frac{\left(\mu_k^D + \pi_k^D\left(\frac{\mu_k^D}{\mu_k^D + \upsilon_k^D}\right)\right)}{\sum_{k=1}^{K}\left(\mu_k^D + \pi_k^D\left(\frac{\mu_k^D}{\mu_k^D + \upsilon_k^D}\right)\right)} \tag{5.17}$$

$$\sum_{k=1}^{K} \lambda_k = 1 \tag{5.18}$$

where $\lambda_k$ is the crisp weight of the $k$-th decision-maker.

The aggregated intuitionistic fuzzy comparison judgement matrix can be obtained through the intuitionistic fuzzy weighted averaging (IFWA) operator (Xu and Yager 2006) by Eqs. 5.19, 5.20.

$$\begin{array}{c} \\ C_1 \\ C_2 \\ \vdots \\ C_n \end{array}\begin{array}{cccc} C_1 & C_2 & \cdots & C_n \\ a_{11} & a_{12} & \cdots & a_{1n} \\ a_{21} & a_{22} & \vdots & a_{2n} \\ \vdots & \vdots & \ddots & \vdots \\ a_{n1} & a_{n2} & \cdots & a_{nn} \end{array} \tag{5.19}$$

$$\begin{aligned} a_{ij} &= IFWA_{\lambda}\left(a_{ij}^{1}, a_{ij}^{2}, \cdots, a_{ij}^{K}\right) \\ &= \left(1-\prod_{k=1}^{K}\left(1-\mu_{a_{ij}^{k}}\right)^{\lambda_k}, \prod_{k=1}^{K}\left(\upsilon_{a_{ij}^{k}}\right)^{\lambda_k}, \prod_{k=1}^{K}\left(1-\mu_{a_{ij}^{k}}\right)^{\lambda_k} - \prod_{k=1}^{K}\left(\upsilon_{a_{ij}^{k}}\right)^{\lambda_k}\right) \end{aligned} \tag{5.20}$$

where $a_{ij} = (\mu_{a_{ij}}, \upsilon_{a_{ij}}, \pi_{a_{ij}})$ represents the element in cell (i, j) in the matrix, and it represents the relative importance/priority of the i-th criterion $C_i$ to that of the j-th criterion $C_j$.

**Step 4**: Determine the product of the elements in each row and the value of $\rho$ with respect to each factor.

The product of the elements in each row can be determined by Eq. 5.21

$$\begin{aligned} r_j &= \bigotimes_{i=1}^{n} a_{ij} = \bigotimes_{i=1}^{n}\left(\mu_{a_{ij}}, \upsilon_{a_{ij}}, \pi_{a_{ij}}\right) \\ &= \left(\prod_{i=1}^{n}\mu_{a_{ij}}, \prod_{i=1}^{n}\left(1-\upsilon_{a_{ij}}\right), 1-\prod_{i=1}^{n}\mu_{a_{ij}} - \prod_{i=1}^{n}\left(1-\upsilon_{a_{ij}}\right)\right) \end{aligned} \tag{5.21}$$

where, $r_j = \left(\mu_{r_j}, \upsilon_{r_j}, \pi_{r_j}\right)$ is the fuzzy geometric mean referring to the j-th criterion.

Then, the value of $\rho$ with respect to each criterion can be determined by Eq. 5.12 or Eq. 5.13, as presented in Eq. 5.22.

$$\rho\left(r_j\right) = \frac{\left(1+\pi_{r_j}\right)\left(1-\mu_{r_j}\right)}{2} \tag{5.22}$$

The smaller the value of $\rho\left(r_j\right)$, the greater $r_j$ in the sense of the amount of positive information included and reliability of information (Xu and Liao, 2014).

**Step 5**: Calculating the weights of the criteria.

The weight of each criterion can then be determined by Eq. 5.23.

$$\omega_j = \frac{1/\rho\left(r_j\right)}{\sum_{j=1}^{n}\left[1/\rho\left(r_j\right)\right]} \tag{5.23}$$

**Table 5.4** Linguistic variables and their corresponding intuitionistic fuzzy numbers (Pramanik and Mukhopadhyaya 2011)

| Linguistic variables | Abbreviation | Intuitionistic fuzzy numbers |
|---|---|---|
| Extreme good | EG | (0.95, 0.05, 0) |
| Very good | VG | (0.85, 0.10, 0.05) |
| Good | G | (0.75, 0.15, 0.10) |
| Medium good | MG | (0.65, 0.25, 0.10) |
| Fair | F | (0.50, 0.40, 0.10) |
| Medium poor | MP | (0.35, 0.55, 0.10) |
| Poor | P | (0.25, 0.65, 0.10) |
| Very poor | VP | (0.15, 0.80, 0.05) |
| Extreme poor | EP | (0.05, 0.95, 0) |

### *5.3.3 Multi-stakeholder Intuitionistic Fuzzy Grey Relational Analysis (GRA)*

The intuitionistic fuzzy grey relational analysis method was presented in this section based on the work of Hou (2010), Wei (2011) and Wei et al. (2011), and there are five steps:

**Step 1**: Using the linguistic to rate the alternatives with respect to each evaluation criterion. Each of the decision-makers will be asked to use the linguistic terms presented in Table 5.4 to rate the alternatives with respect to each evaluation criteria. Let $A_1, A_2, \cdots, A_m$ be m alternatives and $C_1, C_2, \cdots, C_n$ be n evaluation criteria, each of the decision-maker will use the linguistic terms to rate the alternative according to its performance with respect to the evaluation criteria.

**Step 2**: Establishing the decision-making matrix by using the intuitionistic fuzzy numbers. The decision-making matrix can be determined by transforming the linguistic variables determined in step 1 into intuitionistic fuzzy numbers according to Table 5.4. The intuitionistic fuzzy decision-making matrix can then be determined, as presented in Eq. 5.24.

$$D^k = \left\{x_{ij}^k\right\}_{m\times n} = \begin{array}{c} \\ A_1 \\ A_2 \\ \vdots \\ A_m \end{array} \begin{array}{cccc} C_1 & C_2 & \cdots & C_n \\ \left(\mu_{11}^{x,k}, \upsilon_{11}^{x,k}, \pi_{11}^{x,k}\right) & \left(\mu_{12}^{x,k}, \upsilon_{12}^{x,k}, \pi_{12}^{x,k}\right) & \cdots & \left(\mu_{1n}^{x,k}, \upsilon_{1n}^{x,k}, \pi_{1n}^{x,k}\right) \\ \left(\mu_{21}^{x,k}, \upsilon_{21}^{x,k}, \pi_{21}^{x,k}\right) & \left(\mu_{22}^{x,k}, \upsilon_{22}^{x,k}, \pi_{22}^{x,k}\right) & \vdots & \left(\mu_{2n}^{x,k}, \upsilon_{2n}^{x,k}, \pi_{2n}^{x,k}\right) \\ \vdots & \vdots & \ddots & \vdots \\ \left(\mu_{m1}^{x,k}, \upsilon_{m1}^{x,k}, \pi_{m1}^{x,k}\right) & \left(\mu_{m2}^{x,k}, \upsilon_{m2}^{x,k}, \pi_{m2}^{x,k}\right) & \cdots & \left(\mu_{mn}^{x,k}, \upsilon_{mn}^{x,k}, \pi_{mn}^{x,k}\right) \end{array} \tag{5.24}$$

where $x_{ij}^{k} = (\mu_{ij}^{x,k}, \upsilon_{ij}^{x,k}, \pi_{ij}^{x,k})$ is the element of cell (i, j) in the intuitionistic fuzzy decision-making matrix $D^k$ which denotes the performance of the $i$-th alternative with respect to the $j$-th criterion by the $k$-th decision-maker.

Note that there are total K decision-makers, thus, K intuitionistic fuzzy decision-making matrices can be obtained in this step.

**Step 3**: Obtaining the aggregated decision-making matrix. As there are multiple decision-makers participating the decision-making process, the opinions of all the decision-makers should be incorporated in the decision-making matrix. Accordingly, the K decision-making matrix was aggregated into a matrix to address this. The aggregated intuitionistic fuzzy decision-making matrices can be obtained by Eqs. 5.25–5.28.

$$D = \{x_{ij}\}_{m\times n} = \begin{array}{c} \\ A_1 \\ A_2 \\ \vdots \\ A_m \end{array} \begin{array}{cccc} C_1 & C_2 & \cdots & C_n \\ (\mu_{11}^{x}, \upsilon_{11}^{x}, \pi_{11}^{x}) & (\mu_{12}^{x}, \upsilon_{12}^{x}, \pi_{12}^{x}) & \cdots & (\mu_{1n}^{x}, \upsilon_{1n}^{x}, \pi_{1n}^{x}) \\ (\mu_{21}^{x}, \upsilon_{21}^{x}, \pi_{21}^{x}) & (\mu_{22}^{x}, \upsilon_{22}^{x}, \pi_{22}^{x}) & \vdots & (\mu_{2n}^{x}, \upsilon_{2n}^{x}, \pi_{2n}^{x}) \\ \vdots & \vdots & \ddots & \vdots \\ (\mu_{m1}^{x}, \upsilon_{m1}^{x}, \pi_{m1}^{x}) & (\mu_{m2}^{x}, \upsilon_{m2}^{x}, \pi_{m2}^{x}) & \cdots & (\mu_{mn}^{x}, \upsilon_{mn}^{x}, \pi_{mn}^{x}) \end{array} \tag{5.25}$$

$$\mu_{ij}^{x} = 1 - \prod_{k=1}^{K} \left(1 - \mu_{ij}^{x,k}\right)^{\lambda_k} \tag{5.26}$$

$$\upsilon_{ij}^{x} = \prod_{k=1}^{K} \left(\upsilon_{ij}^{x,k}\right)^{\lambda_k} \tag{5.27}$$

$$\pi_{ij}^{x} = \prod_{k=1}^{K} \left(1 - \mu_{ij}^{x,k}\right)^{\lambda_k} - \prod_{k=1}^{K} \left(\upsilon_{ij}^{x,k}\right)^{\lambda_k} \tag{5.28}$$

**Step 4**: Determining the reference solutions (RS). The reference solutions are the positive ideal solutions which satisfy Eqs. 5.29, 5.30.

$$r^{+} = \{r_1^{+}, r_2^{+}, \cdots, r_n^{+}\} \tag{5.29}$$

$$r_j^{+} = \left(\mu_j^{+}, \upsilon_j^{+}, \pi_j^{+}\right) = \left(\max_i \mu_{ij}^{x}, \min_i \upsilon_{ij}^{x}, 1 - \max_i \mu_{ij}^{x} - \min_i \upsilon_{ij}^{x}\right) \tag{5.30}$$

**Step 5**: Calculating the grey relational coefficient (Kuo et al. 2008). The grey relational coefficient of each alternative from the reference solutions with respect to each criterion can be determined by Eq. 5.31. It is worth pointing out that the normalized Hamming distance between the two intuitionistic fuzzy numbers (Grzegorzewski 2004) was used to determine the grey relational coefficient in this study (see Eq. 5.32).

$$\xi_{ij} = \frac{\min_i \min_j d\left(x_{ij}, r_j^+\right) + \rho \max_i \max_j d\left(x_{ij}, r_j^+\right)}{d\left(x_{ij}, r_j^+\right) + \rho \max_i \max_j d\left(x_{ij}, r_j^+\right)} \quad (5.31)$$

$$d\left(x_{ij}, r_j^+\right) = \frac{1}{2}\left(\left|\mu_{ij}^x - \mu_j^+\right| + \left|\upsilon_{ij}^x - \upsilon_j^+\right|\right) \quad (5.32)$$

where $\xi_{ij}$ represents the grey relational coefficient between $x_{ij}$ and $r_j^+$, $d\left(x_{ij}, r_j^+\right)$ is the normalized Hamming distance between the two intuitionistic fuzzy numbers ($x_{ij}$ and $r_j^+$), and $\rho$ represents the identification coefficient, and it usually takes the value of 0.5.

The larger the value of $\xi_{ij}$ (in other words, the larger the grey relational coefficient between $x_{ij}$ and $r_j^+$), the closer $x_{ij}$ to $r_j^+$ (Kuo et al. 2008).

**Step 6**: Calculating the integrated priority degree (grey relational grade) of each alternative (Wei 2011; Ren 2018).

$$\chi_i = \sum_{j=1}^{n} \omega_j \xi_{ij} \quad (5.33)$$

where, $\chi_i$ represents the integrated degree (grey relational grade) of the $i$-th alternative, and $\omega_j$ represents the weight of the $j$-th criterion.

The ranking of these alternatives can be determined based on the rule that the higher the integrated degree, the better the alternative will be. Accordingly, the one having the largest integrated degree can be recognized as the best alternative.

## 5.4 Case Study

Four pathways for the treatment of the sludge in Chongqing (China) were studied by the proposed method, and they are sludge for landscaping, sludge for landfilling, sludge for incineration, and sludge for cement production, the four pathways have been introduced as follows (Meng 2015):

**Sludge for landscaping ($A_1$)**: this scenario for sludge treatment has five steps, including concentration, anaerobic digestion, dewatering, drying, and landscaping;

**Sludge for landfilling ($A_2$)**: this scenario consists of four steps, including concentration, dewatering, lime stabilization, and sanitary landfilling;

**Sludge for incineration ($A_3$)**: there are also four steps in this scenario, and they are concentration, dewatering, drying, and incineration;

**Sludge for cement production ($A_4$)**: this scenario includes four steps, namely, concentration, dewatering, drying, and mixed burning for cement production.

Three groups of representative stakeholders were invited to participate in the sustainability assessment process, and they are scholars and engineers (S#1), administrators and managers (S#2), local residents (S#3). The relative role of the three groups of stakeholders were recognized as "important", "medium", and "medium" which correspond to (0.75, 0.20, 0.05), (0.50, 0.40, 0.10), and (0.50, 0.40, 0.10), respectively. Then, the weight of the role of each stakeholder group can be determined according to Eq. 5.17, and the results were presented in Eqs. 5.33–5.35.

$$\lambda_1 = \frac{0.75 + 0.05\left(\frac{0.75}{0.75+0.20}\right)}{0.75 + 0.05\left(\frac{0.75}{0.75+0.20}\right) + 0.50 + 0.10\left(\frac{0.50}{0.50+0.40}\right) + 0.50 + 0.10\left(\frac{0.50}{0.50+0.40}\right)} = 0.4154 \tag{5.33}$$

$$\lambda_2 = \frac{0.50 + 0.10\left(\frac{0.50}{0.50+0.40}\right)}{0.75 + 0.05\left(\frac{0.75}{0.75+0.20}\right) + 0.50 + 0.10\left(\frac{0.50}{0.50+0.40}\right) + 0.50 + 0.10\left(\frac{0.50}{0.50+0.40}\right)} = 0.2923 \tag{5.34}$$

$$\lambda_3 = \frac{0.50 + 0.10\left(\frac{0.50}{0.50+0.40}\right)}{0.75 + 0.05\left(\frac{0.75}{0.75+0.20}\right) + 0.50 + 0.10\left(\frac{0.50}{0.50+0.40}\right) + 0.50 + 0.10\left(\frac{0.50}{0.50+0.40}\right)} = 0.2923 \tag{5.35}$$

Therefore, the relative weights of the three groups of stakeholders are 0.4154, 0.2923, and 0.2923, respectively.

Each group of stakeholders held a seminar to establish the comparison judgment matrices for determining the weights of the four categories and that of the criteria in each category, as well as rating the alternative pathways for sludge treatment with respect to each criterion. The intuitionistic fuzzy comparison matrices determined by the three groups of stakeholders for calculating the weights of the four categories were presented in Table 5.5.

According to Eqs. 5.19, 5.20, the aggregated intuitionistic fuzzy comparison judgement matrix can be calculated. For instance, cell (1, 2) of Table 5.6 which represents the relative priority of the economic category to the environmental category can be determined by Eq. 5.36.

$$\begin{aligned} a_{12} &= IFWA_\lambda\left(\alpha_{12}^1, \alpha_{12}^2, \alpha_{12}^3\right) \\ &= \left(1 - \prod_{k=1}^{3}\left(1 - \mu_{\alpha_{12}^k}\right)^{\lambda_k}, \prod_{k=1}^{3}\left(\upsilon_{\alpha_{12}^k}\right)^{\lambda_k}, \prod_{k=1}^{3}\left(1 - \mu_{\alpha_{12}^k}\right)^{\lambda_k} - \prod_{k=1}^{3}\left(\upsilon_{\alpha_{12}^k}\right)^{\lambda_k}\right) \end{aligned}$$

**Table 5.5** The intuitionistic fuzzy comparison matrices determined by the three groups of stakeholders for calculating the weights of the four categories

| | Economic | Environmental | Social | Technological |
|---|---|---|---|---|
| *S#1* | | | | |
| Economic | (0.02, 0.18, 0.80) | (0.06, 0.23, 0.70) | (0.22, 0.28, 0.50) | (0.27, 0.13, 0.60) |
| Environmental | (0.23, 0.06, 0.70) | (0.02, 0.18, 0.80) | (0.13, 0.27, 0.60) | (0.23, 0.47, 0.30) |
| Social | (0.28, 0.22, 0.50) | (0.27, 0.13, 0.60) | (0.02, 0.18, 0.80) | (0.18, 0.62, 0.20) |
| Technological | (0.13, 0.27, 0.60) | (0.47, 0.23, 0.30) | (0.62, 0.18, 0.20) | (0.02, 0.18, 0.80) |
| *S#2* | | | | |
| Economic | (0.02, 0.18, 0.80) | (0.02, 0.18, 0.80) | (0.06, 0.23, 0.70) | (0.23, 0.06, 0.70) |
| Environmental | (0.02, 0.18, 0.80) | (0.02, 0.18, 0.80) | (0.22, 0.28, 0.50) | (0.27, 0.33, 0.40) |
| Social | (0.23, 0.06, 0.70) | (0.28, 0.22, 0.50) | (0.02, 0.18, 0.80) | (0.23, 0.47, 0.30) |
| Technological | (0.06, 0.23, 0.70) | (0.33, 0.27, 0.40) | (0.47, 0.23, 0.30) | (0.02, 0.18, 0.80) |
| *S#3* | | | | |
| Economic | (0.02, 0.18, 0.80) | (0.13, 0.27, 0.60) | (0.22, 0.28, 0.50) | (0.27, 0.33, 0.40) |
| Environmental | (0.27, 0.13, 0.60) | (0.02, 0.18, 0.80) | (0.33, 0.27, 0.40) | (0.23, 0.47, 0.30) |
| Social | (0.28, 0.22, 0.50) | (0.27, 0.33, 0.40) | (0.02, 0.18, 0.80) | (0.23, 0.47, 0.30) |
| Technological | (0.33, 0.27, 0.40) | (0.47, 0.23, 0.30) | (0.47, 0.23, 0.30) | (0.02, 0.18, 0.80) |

**Table 5.6** The aggregated intuitionistic fuzzy comparison judgment matrix about the four categories

| | Economic (EC) | Environmental (EN) | Social (S) | Technological (T) |
|---|---|---|---|---|
| EC | (0.02, 0.18, 0.80) | (0.0698, 0.2244, 0.7059) | (0.1763, 0.2644, 0.5594) | (0.2585, 0.1362, 0.60530) |
| EN | (0.1865, 0.1037, 0.7098) | (0.02, 0.18, 0.80) | (0.2193, 0.2729, 0.5078) | (0.2419, 0.4238, 0.3342) |
| S | (0.2657, 0.1505, 0.5838) | (0.2729, 0.1991, 0.5280) | (0.02, 0.18, 0.80) | (0.2096, 0.5273, 0.2631) |
| T | (0.1755, 0.2576, 0.5668) | (0.4324, 0.2410, 0.3265) | (0.5384, 0.2077, 0.2539) | (0.02, 0.18, 0.80) |

$$
\begin{aligned}
&= \begin{pmatrix} 1-(1-0.06)^{0.4154}(1-0.02)^{0.2923}(1-0.13)^{0.2923},(0.23)^{0.4154} \\ (0.18)^{0.2923}(0.27)^{0.2923},(1-0.06)^{0.4154}(1-0.02)^{0.2923} \\ (1-0.13)^{0.2923}-(0.23)^{0.4154}(0.18)^{0.2923}(0.27)^{0.2923} \end{pmatrix} \\
&= (0.0698,0.2244,0.7059) \qquad (5.36)
\end{aligned}
$$

In a similar way, other elements in the aggregated intuitionistic fuzzy comparison judgement matrix can also be determined, and the results were summarized in Table 5.6.

The product of the elements in each row can be determined, and the results were presented in Eq. 5.37.

$$\begin{vmatrix} 6.36E-05 & 0.4042 & 0.5958 \\ 1.98E-04 & 0.3079 & 0.6919 \\ 3.04E-04 & 0.2637 & 0.7360 \\ 8.17E-04 & 0.3660 & 0.6331 \end{vmatrix} \quad (5.37)$$

Then, the value of $\rho$ with respect to each category can be determined by Eq. 5.22, and they are 0.7978, 0.8458, 0.8677, and 0.8159, respectively. Finally, the weights of the four categories can also be determined by Eq. 5.13, and they are 0.2605, 0.2457, 0.2394, and 0.2544, respectively.

Similarly, the weights of the criteria in each category can also be determined, and the results were presented in Tables 5.7, 5.8, 5.9 and 5.10.

After determining the relative importance (weights) of the four categories and that of the criteria in each category, the global weights of the criteria can also be determined. For instance, the global weights of capital cost in economic aspect can be determined by determining the product of the weight of economic category and that of the local weight of capital cost in economic category, namely $0.2605 \times 0.3838 = 0.1000$, thus, the global weight of capital cost is 0.1000. In a similar way, the global weights of other criteria can also be determined, and the results were presented in Table 5.11.

After determining the weights of the criteria, the intuitionistic fuzzy grey relational analysis method was used to rank the four pathways for sludge treatment according to their sustainability. The three groups of decision-makers were firstly

**Table 5.7** The weights of the criteria in the economic aspect

| | Capital cost | Operation cost | Benefit |
|---|---|---|---|
| *S#1* | | | |
| Capital cost | (0.02, 0.18, 0.80) | (0.62, 0.18, 0.20) | (0.80, 0.10, 0.10) |
| Operation cost | (0.18, 0.62, 0.20) | (0.02, 0.18, 0.80) | (0.33, 0.27, 0.40) |
| Benefit | (0.10, 0.80, 0.10) | (0.27, 0.33, 0.40) | (0.02, 0.18, 0.80) |
| *S#2* | | | |
| Capital cost | (0.02, 0.18, 0.80) | (0.47, 0.23, 0.30) | (0.62, 0.18, 0.20) |
| Operation cost | (0.23, 0.47, 0.30) | (0.02, 0.18, 0.80) | (0.22, 0.28, 0.50) |
| Benefit | (0.18, 0.62, 0.20) | (0.28, 0.22, 0.50) | (0.02, 0.18, 0.80) |
| *S#3* | | | |
| Capital cost | (0.02, 0.18, 0.80) | (0.47, 0.23, 0.30) | (0.80, 0.10, 0.10) |
| Operation cost | (0.23, 0.47, 0.30) | (0.02, 0.18, 0.80) | (0.33, 0.27, 0.40) |
| Benefit | (0.10, 0.80, 0.10) | (0.27, 0.33, 0.40) | (0.02, 0.18, 0.80) |
| Weights | 0.3838 | 0.3195 | 0.2967 |

**Table 5.8** The weights of the criteria in the environmental aspect

| | $NO_x$ | $SO_2$ |
|---|---|---|
| *S#1* | | |
| $NO_x$ | (0.02, 0.18, 0.80) | (0.23,0.47, 0.30) |
| $SO_2$ | (0.47, 0.23, 0.30) | (0.02,0.18,0.80) |
| *S#2* | | |
| $NO_x$ | (0.02, 0.18, 0.80) | (0.27, 0.33, 0.40) |
| $SO_2$ | (0.33, 0.27, 0.40) | (0.02, 0.18, 0.80) |
| *S#3* | | |
| $NO_x$ | (0.22, 0.28, 0.50) | (0.28, 0.22, 0.50) |
| $SO_2$ | (0.33, 0.27, 0.40) | (0.02, 0.18, 0.80) |
| Weights | 0.7417 | 0.2583 |

**Table 5.9** The weights of the criteria in the social aspect

| | Added jobs | Compliance with the policy |
|---|---|---|
| *S#1* | | |
| Added jobs | (0.02, 0.18, 0.80) | (0.18, 0.62, 0.20) |
| Compliance with the policy | (0.62, 0.18, 0.20) | (0.02, 0.18, 0.80) |
| *S#2* | | |
| Added jobs | (0.02, 0.18, 0.80) | (0.27, 0.33, 0.40) |
| Compliance with the policy | (0.33, 0.27, 0.40) | (0.02, 0.18, 0.80) |
| *S#3* | | |
| Added jobs | (0.22, 0.28, 0.50) | (0.18, 0.62, 0.20) |
| Compliance with the policy | (0.62, 0.18, 0.20) | (0.02, 0.18, 0.80) |
| Weights | 0.4537 | 0.5463 |

**Table 5.10** The weights of the criteria in the technological aspect

| | Technology maturity | Management requirement |
|---|---|---|
| *S#1* | | |
| Technology maturity | (0.02, 0.18, 0.80) | (0.47, 0.23, 0.30) |
| Management requirement | (0.23, 0.47, 0.30) | (0.02, 0.18, 0.80) |
| *S#2* | | |
| Technology maturity | (0.02, 0.18, 0.80) | (0.80, 0.10, 0.10) |
| Management requirement | (0.10, 0.80, 0.10) | (0.02, 0.18, 0.80) |
| *S#3* | | |
| Technology maturity | (0.22, 0.28, 0.50) | (0.62, 0.18, 0.20) |
| Management requirement | (0.18, 0.62, 0.20) | (0.02, 0.18, 0.80) |
| Weights | 0.5517 | 0.4483 |

**Table 5.11** The global weights of the criteria for sustainability assessment

| Criteria | $EC_1$ | $EC_2$ | $EC_3$ | $EN_1$ | $EN_2$ | $S_1$ | $S_2$ | $T_1$ | $T_2$ |
|---|---|---|---|---|---|---|---|---|---|
| Weights | 0.1000 | 0.0832 | 0.0773 | 0.1822 | 0.0635 | 0.1086 | 0.1308 | 0.1404 | 0.1140 |

**Table 5.12** The results for rating the alternatives using the linguistic variables determined by S#1

| | $A_1$ | $A_2$ | $A_3$ | $A_4$ |
|---|---|---|---|---|
| $EC_1$ | MP | G | F | MG |
| $EC_2$ | MP | F | VP | VG |
| $EC_3$ | F | MG | F | MG |
| $EN_1$ | G | MP | EP | MP |
| $EN_2$ | MG | P | VP | P |
| $S_1$ | MP | P | G | F |
| $S_2$ | VG | MP | G | F |
| $T_1$ | EG | VG | F | MP |
| $T_2$ | MG | P | F | P |

asked to use the linguistic variables presented in Table 5.4 to rate the four alternatives with respect to each evaluation criterion, and the results determined by S#1 were presented in Table 5.12 (the results for rating the alternatives using the linguistic variables determined by the other two groups of decision-makers were presented in the Appendix).

According to Table 5.4, the linguistic variables presented in Table 5.12 can be transformed into intuitionistic fuzzy numbers. For instance, "G" in cell (1,2) can be transformed into $(0.75, 0.15, 0.10)$. In a similar way, the other linguistic variables can also be transformed into intuitionistic fuzzy numbers (see Table 5.13). The intuitionistic fuzzy decision-making matrices determined by S#2 and S#3 were presented in the Appendix.

**Table 5.13** The intuitionistic fuzzy decision-making matrix determined by S#1

| | $A_1$ | $A_2$ | $A_3$ | $A_4$ |
|---|---|---|---|---|
| $EC_1$ | (0.35, 0.55, 0.10) | (0.75, 0.15, 0.10) | (0.50, 0.40, 0.10) | (0.65, 0.25, 0.10) |
| $EC_2$ | (0.35, 0.55, 0.10) | (0.50, 0.40, 0.10) | (0.15, 0.80, 0.05) | (0.85, 0.10, 0.05) |
| $EC_3$ | (0.50, 0.40, 0.10) | (0.65, 0.25, 0.10) | (0.50, 0.40, 0.10) | (0.65, 0.25, 0.10) |
| $EN_1$ | (0.75, 0.15, 0.10) | (0.35, 0.55, 0.10) | (0.05, 0.95, 0) | (0.35, 0.55, 0.10) |
| $EN_2$ | (0.65, 0.25, 0.10) | (0.25, 0.65, 0.10) | (0.15, 0.80, 0.05) | (0.25, 0.65, 0.10) |
| $S_1$ | (0.35, 0.55, 0.10) | (0.25, 0.65, 0.10) | (0.75, 0.15, 0.10) | (0.50, 0.40, 0.10) |
| $S_2$ | (0.85, 0.10, 0.05) | (0.35, 0.55, 0.10) | (0.75, 0.15, 0.10) | (0.50, 0.40, 0.10) |
| $T_1$ | (0.95, 0.05, 0) | (0.85, 0.10, 0.05) | (0.50, 0.40, 0.10) | (0.35, 0.55, 0.10) |
| $T_2$ | (0.65, 0.25, 0.10) | (0.25, 0.65, 0.10) | (0.50, 0.40, 0.10) | (0.25, 0.65, 0.10) |

**Table 5.14** The aggregated decision-making matrix determined

| | $A_1$ | $A_2$ | $A_3$ | $A_4$ | RS |
|---|---|---|---|---|---|
| $EC_1$ | (0.3222, 0.5775, 0.1002) | (0.7847, 0.1332, 0.0821) | (0.0821, 0.3663, 0.1025) | (0.6828, 0.2153, 0.1019) | (0.7847, 0.1332, 0.0821) |
| $EC_2$ | (0.3890, 0.5011, 0.1009) | (0.5495, 0.3487, 0.1018) | (0.2100, 0.7086, 0.0815) | (0.7769, 0.1472, 0.0759) | (0.7769, 0.1472, 0.0759) |
| $EC_3$ | (0.4171, 0.4818, 0.1010) | (0.7125, 0.1855, 0.1020) | (0.3696, 0.5376, 0.0928) | (0.6115, 0.2868, 0.1016) | (0.7125, 0.1855, 0.1020) |
| $EN_1$ | (0.6622, 0.2320, 0.1058) | (0.3222, 0.5775, 0.1002) | (0.0804, 0.9035, 0.0162) | (0.3980, 0.5011, 0.1009) | (0.6622, 0.2320, 0.1058) |
| $EN_2$ | (0.6115, 0.2868, 0.1016) | (0.3611, 0.5371, 0.1018) | (0.2423, 0.6748, 0.0829) | (0.3611, 0.5371, 0.1018) | (0.6115, 0.2868, 0.1016) |
| $S_1$ | (0.3222, 0.5775, 0.1002) | (0.2221, 0.6907, 0.0873) | (0.7624, 0.1547, 0.0829) | (0.3922, 0.5060, 0.1018) | (0.7624, 0.1547, 0.0829) |
| $S_2$ | (0.7769, 0.1472, 0.0759) | (0.2933, 0.6064, 0.1003) | (0.6957, 0.2022, 0.1021) | (0.4601, 0.4390, 0.1008) | (0.7769, 0.1472, 0.0759) |
| $T_1$ | (0.8897, 0.0844, 0.0259) | (0.7769, 0.1472, 0.0759) | (0.5495, 0.3487, 0.1018) | (0.3980, 0.5011, 0.1009) | (0.8897, 0.0844, 0.0259) |
| $T_2$ | (0.6479, 0.2470, 0.1050) | (0.2807, 0.6190, 0.1003) | (0.4601, 0.4390, 0.1008) | (0.2221, 0.6907, 0.0873) | (0.6479, 0.2470, 0.1050) |

The three intuitionistic fuzzy decision-making matrices were aggregated into the aggregated decision-making matrix according to Eqs. 5.25–5.28 (see Table 5.14).

Subsequently, the reference solutions can be determined by Eqs. 5.29, 5.30, and the results were also presented in Table 5.14. Then, grey relational coefficient of each alternative from the reference solutions can be determined according to Eqs. 5.31, 5.32, and the results were presented in Table 5.15. Finally, the integrated priority degree of each alternative can be determined by Eq. 5.33, and the results were presented in Fig. 5.1.

It is obvious that sludge for landscaping ($A_1$) has been recognized as the most sustainable pathway for sludge treatment, follows by sludge for landfilling ($A_2$), sludge for cement production ($A_4$), and sludge for incineration ($A_3$). There are many reasons for sludge for landscaping ($A_1$) being the most sustainable scenario among these four alternative pathways for sludge treatment though the capital cost and operation cost are very high: (i) sludge for landscaping has considerable economic benefits; (ii) high compliance with the policy; (iii) relatively lower environmental impacts, less

**Table 5.15** The grey relational coefficient of each alternative from the reference solutions

| | $A_1$ | $A_2$ | $A_3$ | $A_4$ |
|---|---|---|---|---|
| $EC_1$ | 0.3652 | 1.0000 | 0.3900 | 0.7406 |
| $EC_2$ | 0.4161 | 0.5497 | 0.3160 | 1.0000 |
| $EC_3$ | 0.4691 | 1.0000 | 0.4291 | 0.7220 |
| $EN_1$ | 1.0000 | 0.4325 | 0.2937 | 0.4951 |
| $EN_2$ | 1.0000 | 0.5110 | 0.4082 | 0.5110 |
| $S_1$ | 0.3768 | 0.3263 | 1.0000 | 0.4199 |
| $S_2$ | 1.0000 | 0.3562 | 0.7943 | 0.4620 |
| $T_1$ | 1.0000 | 0.7496 | 0.4637 | 0.3648 |
| $T_2$ | 1.0000 | 0.4140 | 0.5797 | 0.3751 |

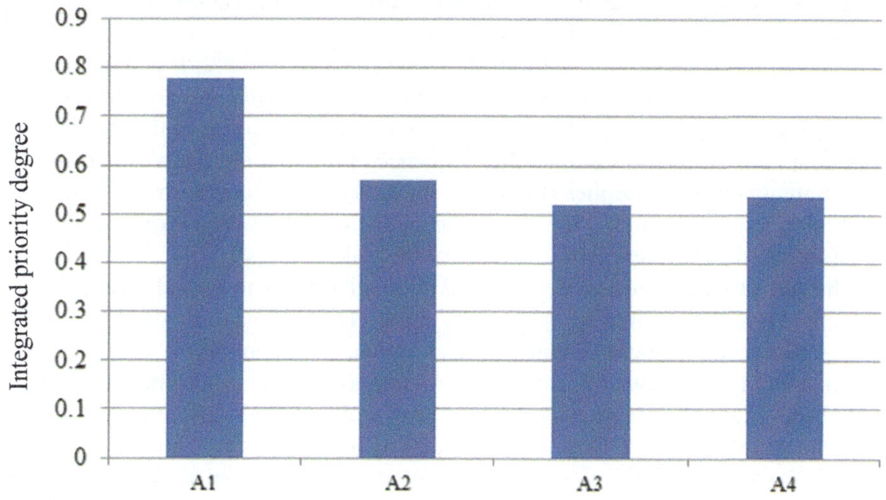


**Fig. 5.1** The integrated priority degree

emissions, and great ecological benefits; and (iv) high technology maturity and low management requirement.

## 5.5 Discussion

The sum weighted method (SWM) method through the intuitionistic fuzzy weighted averaging (IFWA) operator (Xu 2007) was also employed to validate the results determined by the intuitionistic fuzzy grey relational analysis method based on the aggregated intuitionistic fuzzy decision-making matrix and the weights determined by intuitionistic fuzzy AHP, and the comparison of the results were presented in Table 5.16.

**Table 5.16** The comparison of the results determined by

| | $A_1$ | $A_2$ | $A_3$ | $A_4$ |
|---|---|---|---|---|
| Ranking by the intuitionistic fuzzy GRA | 1 | 2 | 4 | 3 |
| Ranking by the WSM method | 1 | 2 | 4 | 3 |

The results determined by these two methods are the same. To some extent, it demonstrates that the sustainability sequence of the four alternative pathways for sludge treatment is correct.

Sensitivity analysis was employed to study the influences of the weights of the criteria on the results, and the ten cases presented in Table 5.17 have been investigated to compare with the base case the weights determined by intuitionistic fuzzy AHP. Note that all the criteria were recognized as equal importance in Case 2, and a dominant weight was assigned to one criterion and the other criteria were recognized as equally important in the other nine cases.

The results of sensitivity analysis were presented in Fig. 5.2. It is apparent that the integrated priority degrees of the alternatives were sensitive to the weights of the criteria for sustainability assessment. In other words, the weights of the criteria have significant effects on the sustainability sequence of the four alternative pathways for sludge treatment. The implies that the accurate determination of the weights is of vital importance for ranking the alternative pathways for sludge treatment correctly according to the preferences/willingness of the stakeholders.

The sustainability sequence of the four alternative pathways for sludge treatment from the most sustainable to the least is sludge for landscaping, sludge for landfilling, sludge for cement production, and sludge for incineration. However, it is worth pointing out that the priority sequence may change when the stakeholders have been changed, because (i) different stakeholders have different preferences, thus, the weights of the criteria for sustainability assessment which have significant impacts on the final ranking may also change; (ii) the intuitionistic fuzzy decision-making matrices determined by different stakeholders may also be different due to the difference in opinions and perceptions.

## 5.6 Conclusion

This study aims at developing a multi-actor multi-criteria decision supporting method for help the stakeholders to assess the sustainability of the alternative pathways for sludge treatment, and the model based on intuitionistic fuzzy set theory was presented in this study. An evaluation criteria system which consists of nine criteria in four categories for sustainability assessment was developed. The intuitionistic fuzzy AHP which can incorporate the opinions of multiple stakeholders and address the vagueness and ambiguity in the judgments of the stakeholders was developed to determine the weights of the criteria for sustainability assessment of the technologies

**Table 5.17** The weights of the criteria set in differert cases

| | Case 1 | Case 2 | Case 3 | Case 4 | Case 5 | Case 6 | Case 7 | Case 8 | Case 9 | Case 10 |
|---|---|---|---|---|---|---|---|---|---|---|
| $EC_1$ | 0.1111 | 0.36 | 0.08 | 0.08 | 0.08 | 0.08 | 0.08 | 0.08 | 0.08 | 0.08 |
| $EC_2$ | 0.1111 | 0.08 | 0.36 | 0.08 | 0.08 | 0.08 | 0.08 | 0.08 | 0.08 | 0.08 |
| $EC_3$ | 0.1111 | 0.08 | 0.08 | 0.36 | 0.08 | 0.08 | 0.08 | 0.08 | 0.08 | 0.08 |
| $EN_1$ | 0.1111 | 0.08 | 0.08 | 0.08 | 0.36 | 0.08 | 0.08 | 0.08 | 0.08 | 0.08 |
| $EN_2$ | 0.1111 | 0.08 | 0.08 | 0.08 | 0.08 | 0.36 | 0.08 | 0.08 | 0.08 | 0.08 |
| $S_1$ | 0.1111 | 0.08 | 0.08 | 0.08 | 0.08 | 0.08 | 0.36 | 0.08 | 0.08 | 0.08 |
| $S_2$ | 0.1111 | 0.08 | 0.08 | 0.08 | 0.08 | 0.08 | 0.08 | 0.36 | 0.08 | 0.08 |
| $T_1$ | 0.1111 | 0.08 | 0.08 | 0.08 | 0.08 | 0.08 | 0.08 | 0.08 | 0.36 | 0.08 |
| $T_2$ | 0.1111 | 0.08 | 0.08 | 0.08 | 0.08 | 0.08 | 0.08 | 0.08 | 0.08 | 0.36 |

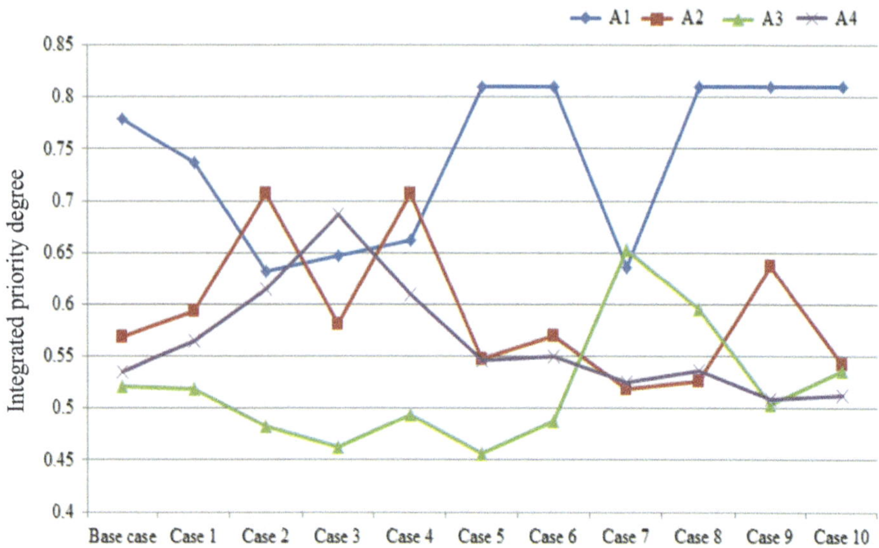


**Fig. 5.2** The results of sensitivity analysis

for sludge treatment. The intuitionistic fuzzy grey relational analysis method was employed to ranking the alternative pathways for sludge treatment according to their comprehensive performances with respect to sustainability. All the proposed methods for sustainability assessment of sludge treatment technologies has the following advantages:

(i) Multiple stakeholders can participate in the sustainability assessment process, it can incorporate the preference/willingness of different representative groups of decision-makers to achieve group decision-making;
(ii) The multi-stakeholder multi-criteria decision making method based on intuitionistic fuzzy set theory can effectively solve the uncertainties existing in the judgements of the stakeholders;
(iii) The decision-making matrix was established based on the intuitionistic judgments of the stakeholders rather than exact data. Thus, sustainability assessment and prioritization can be carried out even without exact data of the alternatives with respect to the evaluation criteria.

However, decision-making without exact data is a double-edged sword, and it cannot fully use the obtained information, because decision-making was not based on the exact data. For instance, the value of one scenario with respect to capital cost cannot be used even if the stakeholders know the exact data. Therefore, the future work of the authors is to develop a modified multi-actor multi-criteria decision supporting method which can use the useful information (especially the data) for sustainability assessment of the alternative technologies for sludge treatment.

**Acknowledgements** This study was financially supported by the Hong Kong Research Grants Council for Early Career Scheme (Grant No. 25208118).

# Appendix

See Tables 5.18, 5.19, 5.20, 5.21

**Table 5.18** The results for rating the alternatives using the linguistic variables determined by S#2

| | $A_1$ | $A_2$ | $A_3$ | $A_4$ |
|---|---|---|---|---|
| $EC_1$ | P | VG | P | G |
| $EC_2$ | F | MG | P | G |
| $EC_3$ | MP | G | MP | MG |
| $EN_1$ | MG | MP | VP | F |
| $EN_2$ | F | MP | P | MP |
| $S_1$ | P | VP | VG | MP |
| $S_2$ | G | P | MG | MP |
| $T_1$ | VG | G | MG | F |
| $T_2$ | G | P | MP | VP |

**Table 5.19** The results for rating the alternatives using the linguistic variables determined by S#3

| | $A_1$ | $A_2$ | $A_3$ | $A_4$ |
|---|---|---|---|---|
| $EC_1$ | MP | G | P | MG |
| $EC_2$ | MP | F | P | MG |
| $EC_3$ | MP | G | VP | F |
| $EN_1$ | F | P | EP | MP |
| $EN_2$ | MG | F | MP | F |
| $S_1$ | MP | P | MG | P |
| $S_2$ | MG | P | MG | F |
| $T_1$ | G | MG | F | MP |
| $T_2$ | F | MP | F | P |

**Table 5.20** The intuitionistic fuzzy decision-making matrix determined by S#2

| | $A_1$ | $A_2$ | $A_3$ | $A_4$ |
|---|---|---|---|---|
| $EC_1$ | (0.25, 0.65, 0.10) | (0.85, 0.10, 0.05) | (0.25, 0.65, 0.10) | (0.75, 0.15, 0.10) |
| $EC_2$ | (0.50, 0.40, 0.10) | (0.65, 0.25, 0.10) | (0.25, 0.65, 0.10) | (0.75, 0.15, 0.10) |
| $EC_3$ | (0.35, 0.55, 0.10) | (0.75, 0.15, 0.10) | (0.35, 0.55, 0.10) | (0.65, 0.25, 0.10) |
| $EN_1$ | (0.65, 0.25, 0.10) | (0.35, 0.55, 0.10) | (0.15, 0.80, 0.05) | (0.50, 0.40, 0.10) |
| $EN_2$ | (0.50, 0.40, 0.10) | (0.35, 0.55, 0.10) | (0.25, 0.65, 0.10) | (0.35, 0.55, 0.10) |
| $S_1$ | (0.25, 0.65, 0.10) | (0.15, 0.80, 0.05) | (0.85, 0.10, 0.05) | (0.35, 0.55, 0.10) |
| $S_2$ | (0.75, 0.15, 0.10) | (0.25, 0.65, 0.10) | (0.65, 0.25, 0.10) | (0.35, 0.55, 0.10) |
| $T_1$ | (0.85, 0.10, 0.05) | (0.75, 0.15, 0.10) | (0.65, 0.25, 0.10) | (0.50, 0.40, 0.10) |
| $T_2$ | (0.75, 0.15, 0.10) | (0.25, 0.65, 0.10) | (0.35, 0.55, 0.10) | (0.15, 0.80, 0.05) |

**Table 5.21** The intuitionistic fuzzy decision-making matrix determined by S#3

| | $A_1$ | $A_2$ | $A_3$ | $A_4$ |
|---|---|---|---|---|
| $EC_1$ | (0.35, 0.55, 0.10) | (0.75, 0.15, 0.10) | (0.25, 0.65, 0.10) | (0.65, 0.25, 0.10) |
| $EC_2$ | (0.35, 0.55, 0.10) | (0.50, 0.40, 0.10) | (0.25, 0.65, 0.10) | (0.65, 0.25, 0.10) |
| $EC_3$ | (0.35, 0.55, 0.10) | (0.75, 0.15, 0.10) | (0.15, 0.80, 0.05) | (0.50, 0.40, 0.10) |
| $EN_1$ | (0.50, 0.40, 0.10) | (0.25, 0.65, 0.10) | (0.05, 0.95, 0) | (0.35, 0.55, 0.10) |
| $EN_2$ | (0.65, 0.25, 0.10) | (0.50, 0.40, 0.10) | (0.35, 0.55, 0.10) | (0.50, 0.40, 0.10) |
| $S_1$ | (0.35, 0.55, 0.10) | (0.25, 0.65, 0.10) | (0.65, 0.25, 0.10) | (0.25, 0.65, 0.10) |
| $S_2$ | (0.65, 0.25, 0.10) | (0.25, 0.65, 0.10) | (0.65, 0.25, 0.10) | (0.50, 0.40, 0.10) |
| $T_1$ | (0.75, 0.15, 0.10) | (0.65, 0.25, 0.10) | (0.50, 0.40, 0.10) | (0.35, 0.55, 0.10) |
| $T_2$ | (0.50, 0.40, 0.10) | (0.35, 0.55, 0.10) | (0.50, 0.40, 0.10) | (0.25, 0.65, 0.10) |

# References


L. Abdullah, L. Najib, A new preference scale of intuitionistic fuzzy analytic hierarchy process in multi-criteria decision making problems. J. Intell. Fuzzy Syst. **26**(2), 1039–1049 (2014)

N.S. Akwo, A life cycle assessment of sewage sludge treatment options. Master Thesis in Environmental Management of Aalborg University. (2008), http://projekter.aau.dk/projekter/files/14465524/final_thesis.pdf. Accessed 1 Apr 2019

F.E. Boran, S. Genc, M. Kurt, D. Akay, A multi-criteria intuitionistic fuzzy group decision making for supplier selection with TOPSIS method. Expert Syst. Appl. **36**(8), 11363–11368 (2009)

H. Chen, S.H. Yan, Z.L. Ye, H.J. Meng, Y.G. Zhu, Utilization of urban sewage sludge: Chinese perspectives. Environ. Sci. Pollut. Res. **19**(5), 1454–1463 (2012)

T. Chen, A prioritized aggregation operator-based approach to multiple criteria decision making using interval-valued intuitionistic fuzzy sets: a comparative perspective. Inf. Sci. (Ny) **281**, 97–112 (2014)

E.O. Diemuodeke, S. Hamilton, A. Addo, Multi-criteria assessment of hybrid renewable energy systems for Nigeria's coastline communities. Energy Sustain. Soc. **6**(1), 26 (2016)

O. Eriksson, M. Bisaillon, M. Haraldsson, J. Sundberg, Enhancement of biogas production from food waste and sewage sludge–Environmental and economic life cycle performance. J. Environ. Manage. **175**, 33–39 (2016)

S. Fersi, N. Chtourou, C. Jury, F. Poncelet, Economic analysis of renewable heat and electricity production by sewage sludge digestion—a case study. Int. J. Energy Res. **39**(2), 234–243 (2015)

F. Flocard, D. Ierodiaconou, I.R. Coghlan, Multi-criteria evaluation of wave energy projects on the south-east Australian coast. Renew. Energy **99**, 80–94 (2016)

L. Gigović, D. Pamučar, D. Božanić, S. Ljubojević, Application of the GIS-DANP-MABAC multi-criteria model for selecting the location of wind farms: a case study of Vojvodina, Serbia. Renew. Energy **103**, 501–521 (2017)

P. Grzegorzewski, Distances between intuitionistic fuzzy sets and/or interval-valued fuzzy sets based on the Hausdorff metric. Fuzzy Sets Syst. **148**(2), 319–328 (2004)

B. Haddad, A. Liazid, P. Ferreira, A multi-criteria approach to rank renewables for the Algerian electricity system. Renew. Energy **107**, 462–472 (2017)

A. Hospido, T. Moreira, M. Martín, M. Rigola, G. Feijoo, Environmental evaluation of different treatment processes for sludge from urban wastewater treatments: Anaerobic digestion versus thermal processes. Int. J. Life Cycle Assess. **10**(5), 336–345 (2005)

J. Hou, Grey relational analysis method for multiple attribute decision making in intuitionistic fuzzy setting. J. Converg. Inf. Technol. **5**(10), 194–199 (2010)

Y. Kuo, T. Yang, G.W. Huang, The use of grey relational analysis in solving multiple attribute decision-making problems. Comput. Ind. Eng. **55**(1), 80–93 (2008)

J. Lederer, H. Rechberger, Comparative goal-oriented assessment of conventional and alternative sewage sludge treatment options. Waste Manage. **30**(6), 1043–1056 (2010)

J. Meng, Comprehensive evaluation of urban sludge treatment and disposal program. Master Dissertation of Xi'an University of Architecture and Technology (2015), http://kns.cnki.net/KCMS/detail/detail.aspx?dbcode=CMFD&dbname=CMFD201601&filename=1015994675.nh&v=MTg3NTBoMVQzcVRyV00xRnJDVVJMMmZaT2RwRnk3a1VyL05WRjI2RzdxeEd0ZkxxcEViUElSOGVYMUx1eFlTN0Q. Accessed 1 Apr 2019

G. Pokoo-Aikins, A. Heath, R.A. Mentzer, M.S. Mannan, W.J. Rogers, M.M. El-Halwagi, A multi-criteria approach to screening alternatives for converting sewage sludge to biodiesel. J. Loss Prev. Process Ind. **23**(3), 412–420 (2010)

S. Pramanik, D. Mukhopadhyaya, Grey relational analysis based intuitionistic fuzzy multi-criteria group decision-making approach for teacher selection in higher education. Int. J. Comput. Appl. **34**(10), 21–29 (2011)

J. Ren, Multi-criteria decision making for the prioritization of energy systems under uncertainties after life cycle sustainability assessment. Sustain. Prod. Consum **16**, 45–57 (2018)

J. Ren, H. Liang, L. Dong, Z. Gao, C. He, M. Pan, L. Sun, Sustainable development of sewage sludge-to-energy in China: barriers identification and technologies prioritization. Renew. Sustain. Energ Rev. **67**, 384–396 (2017)

T.L. Saaty, *The Analytic Hierarchy Process* (McGraw-Hill Book Co, New York, 1980)

M. Svanström, M. Fröling, M. Olofsson, M. Lundin, Environmental assessment of supercritical water oxidation of sewage sludge. Resour. Conserv. Recycl. **41**(4), 321–338 (2005)

O. Taylan, D. Kaya, A. Demirbas, An integrated multi attribute decision model for energy efficiency processes in petrochemical industry applying fuzzy set theory. Energy Conv. Manag. **117**, 501–512 (2016)

G.W. Wei, Gray relational analysis method for intuitionistic fuzzy multiple attribute decision making. Expert Syst. Appl. **38**(9), 11671–11677 (2011)

G. Wei, H.J. Wang, R. Lin, X. Zhao, Grey relational analysis method for intuitionistic fuzzy multiple attribute decision making with preference information on alternatives. Int. J. Comput. Intell. Syst. **4**(2), 164–173 (2011)

Z. Xu, Intuitionistic fuzzy aggregation operators. IEEE Trans. Fuzzy Syst. **15**(6), 1179–1187 (2007)

Z. Xu, H. Liao, Intuitionistic fuzzy analytic hierarchy process. IEEE Trans. Fuzzy Syst. **22**(4), 749–761 (2014)

Z. Xu, R. R. Yager, Some geometric aggregation operators based on intuitionistic fuzzy sets. Int. J. Gen Syst. **35**(4), 417–433 (2006)

H.Y. Yap, J.D. Nixon, A multi-criteria analysis of options for energy recovery from municipal solid waste in India and the UK. Waste Manage. **46**, 265–277 (2015)
X. Yu, Z. Xu, Prioritized intuitionistic fuzzy aggregation operators. Inf. Fusion **14**, 108–116 (2013)
X. Yuan, H. Li, C. Zhang, The theory of intuitionistic fuzzy sets based on the intuitionistic fuzzy special sets q. Inf. Sci. (Ny) **277**, 284–298 (2014)
C. Yue, A geometric approach for ranking interval-valued intuitionistic fuzzy numbers with an application to group decision-making. Comput. Ind. Eng. **102**, 233–245 (2016)
L.A. Zadeh, Fuzzy logic—a personal perspective. Fuzzy Sets Syst. **281**, 4–20 (2015)
S.F. Zhang, S.Y. Liu, A GRA-based intuitionistic fuzzy multi-criteria group decision making method for personnel selection. Expert Syst. Appl. **38**(9), 11401–11405 (2011)

The manufacturer's authorised representative in the EU is Springer Nature Customer Service Centre GmbH, Europaplatz 3, 69115 Heidelberg, Germany. If you have any concerns regarding our products, please contact ProductSafety@springernature.com

Printed and bound by CPI Group (UK) Ltd, Croydon, CR0 4YY
30/09/2026
02225575-0007